Plants for Tropical Landscapes

Plants for Tropical Landscapes
A Gardener's Guide

Fred D. Rauch and Paul R. Weissich

University of Hawai'i Press
Honolulu

This guidebook was published with funding assistance
from the Kaulunani Urban Forestry Program. Kaulunani is a
State of Hawaiʻi, Department of Land and Natural Resources,
Division of Forestry and Wildlife program, an equal opportunity
service provider and grantee of the United States Department of
Agriculture Forest Service, Pacific Southwest Region.

© 2000 University of Hawaiʻi Press
All rights reserved
Printed in Singapore

10 09 08 07 06 05   8 7 6 5 4 3

Library of Congress Cataloging-in-Publication Data

Rauch, Fred D. (Fred Donald), 1931–
    Plants for tropical landscapes : a gardener's guide / Fred D.
  Rauch and Paul R. Weissich.
      p.   cm.
  Includes bibliographical references.
  ISBN-13: 978-0-8248-2034-3 (cloth)
  ISBN-10: 0-8248-2034-7 (cloth)
  1. Tropical plants—Hawaii. 2. Landscape plants—Hawaii.
3. Tropical plants. 4. Landscape plants—Tropics. I. Weissich,
Paul R. II. Title
SB407. R37 2000
635.9′523′09969—dc21   99-26677
                                                                            CIP

University of Hawaiʻi Press books are printed on acid-free paper
and meet the guidelines for permanence and durability of the
Council on Library Resources.

Designed by David Alcorn, Alcorn Publication Design,
Red Bluff, California.

Printed by Tien Wah Press (PTE) Limited,
Singapore.

This book is dedicated to Marian and Pat
for their loving patience and forgiveness

Contents

Introduction		ix
1.	Ground Covers	1
2.	Small Shrubs	13
3.	Medium Shrubs	21
4.	Large Shrubs	35
5.	Small Trees	47
6.	Medium Trees	57
7.	Large Trees	71
8.	Vines	77
9.	Bromeliads, Ferns, and Marantas	89
10.	Bamboos, Cycads, and Palms	101

Appendices: Special Use Plants
    A. Landscaping with Native Hawaiian Species   115
    B. Plants for the Xeriscape   116
    C. Plants for the Beach Garden   117
    D. Plants for Hedges, Screens, and Windbreaks   119
    E. House Plants   120
    F. Lanai Plants   121
    G. The Garden at Night   122

Glossary of Terms   125
Suggested Readings   127
Index   129

Cassia x *nealiae* 'Lunalilo Yellow'

Introduction

There is a need for a reference guide to readily available landscape plants for gardens in Hawai'i. This publication provides a photographic guide with brief descriptions of over five hundred tropical plants useful for home landscaping. Basic information relative to cultural requirements and physical characteristics is included. It is hoped that this guide will also be helpful to the student, professional members of the "Green Industry," those in other tropical areas outside of Hawai'i, and to all those interested in tropical plants.

Plant Names

The nomenclature of many tropical plant groups is under constant taxonomic study, resulting in corrective name changes. Scientific names used in this publication are primarily those appearing in the Bishop Museum's *In Gardens of Hawai'i* project. Other taxonomic authority was provided by *Hortus Third*, 1976 (Macmillan Publishing Co.), *The Plant Book*, by Mabberley, 1990 (Cambridge University Press) and, for native Hawaiian plants, the *Manual of the Flowering Plants of Hawai'i* by Wagner, Herbst, and Sohmer, 1990 (Bishop Museum Press). Every effort has been made to obtain the most current plant name information. Any errors are the sole responsibility of the authors. Plants often have many common names throughout the world, which can lead to confusion. An attempt was made to select those most commonly used in Hawai'i and to indicate Hawaiian names where available. For a few species, no common names were found in the literature. We have taken the liberty of suggesting names in these cases.

Plant Selection

Here are some helpful guidelines for the selection of the best plants for your home landscape:
1. Determine the function of the desired plant in your garden, such as a screen, shade tree, or focal point (specimen).
2. Determine the desired characteristics the plant or plants should have, such as height, shape, or flower color.
3. Determine your garden's environmental characteristics, such as rainfall, amount of sunshine, wind exposure, salt air exposure, and soil type.
4. Using this plant guide, match the environmental characteristics with the plant requirements. This will allow you to select and group your plants to create a successful low-maintenance garden tailored to your needs and taste.

Plant Size and Use

The plants in this guide have been grouped by type (such as trees and shrubs) and size for the convenience of the reader. The suggested plant heights are our best estimates of useful sizes under average local conditions and within the life expectancy of the garden. Keep in mind, however, that the growth rate and size of a specific plant can vary with location (owing to variations in rainfall, temperature, soil type, etc.) and, especially, the care provided. Appendices at the end of this guide summarize seven groups of plants of special interest to the gardener. There is also a list of suggested supplemental reading that will enhance the gardener's knowledge of tropical plants. Three groups of plants have been omitted due to space constraints and the fact that ample written information is available from other sources. These are orchids, plants for water gardens, and turf grasses. Publications dealing with these subjects are in the suggested supplemental reading list.

A Word to the Wise Gardener

Plants contain a great variety of chemical and physical properties that may affect their selection for garden use. Many of these have been refined by man to make medicines, dyes, fabrics, and a host of familiar useful products. But some have thorns or irritating hairs, while others produce substances that may cause skin irritations, illness, or fatal poisoning if touched, inhaled, or ingested. Some are more virulent than others.

Susceptibility and reactions vary from one person to another and in accordance with exposure. They are significantly greater in children. All are rendered harmless if recognized and treated with respect and intelligence much as one would, for example, avoid contact with sharp knives, a hot stove, or the claws of an angry cat. A practical rule of thumb is: If you don't know what it is, don't smell it, don't pick it, and above all, don't chew or eat it. Watch the kids. They have been known to chew—even swallow—flowers, leaves, or fruits while playing house or having a tea party, with disastrous results. The list of recommended reading includes sources of information on this subject.

Those plants in the text that may involve one or

more potential problems are given a letter to the right of the common name to indicate the following: (T) plants with thorns or irritating hairs; (S) plants that are known to have caused skin, mouth, throat, or eye irritations; (P) plants with toxic sap or bearing flowers, leaves, or seeds containing toxins. Plants deserving extra care when handling are marked with a plus (+) sign.

Acknowledgements

The authors are indebted to the following for their generous assistance with special problems: Ray Baker (aglaonema, palms); Peter Berg and Susan Ruskin (bamboo); Dr. Richard Criley (ixora); Robert Hirano (Marantaceae, ti); Donald Hodel (chamaedorea palms); Dr. Helen Kennedy (Marantaceae); Dr. Charles Lamoureux (ferns); Dr. Kenneth Leonhardt (cut tropicals); Leland Miyano (cycads and bromels); Lisa Vinzant (bromels), Norman Bezona (palms), Dale H. Sato (ground covers), and David Yearian (ti).

A special mahalo goes out to Jan McEwen and Marian Rauch for computer assistance and to Jim Little for photographic advice. We are deeply appreciative of the cooperation of the Department of Horticulture, University of Hawai'i.

A number of nursery growers have assisted in the identification of some of the newer plants and graciously made their plants available for photographs, including:

    Peter Berg and Susan Ruskin, Quindembo Bamboo Nursery
    Bill Durstan, Leilani Nursery, Inc.
    Patrick McGrath, Hawaii Foliage Exports, Inc.
    Leland Miyano
    Masuo Moriwaki, Moriwaki Nursery
    Glenn Nii, Charles Nii Nursery
    Richard and Judy Nii, R & S Nii Nursery
    Lelan C. Nishek, Kauai Nursery and Landscaping
    Patrick Oka, Oka Nursery and Landscaping
    Sharon Peterson, Sharon's Plants Ltd.
    Pat Sumida, Marugame Nursery
    John Swim, Mokuleia Landscape and Nursery Co.
    Ken & Lisa Vinzant, Olomana Tropicals
    David Yearian, Tis Unlimited

Photo credits:

The photographs used in this publication were taken by Fred D. Rauch, with the exception of the following, which were generously provided by:

    John C. Eveland: *Alyxia oliviformis, Heliotropium anomalum* var. *argenteum, Lipochaeta integrifolia, Odontosoria chinensis, Psilotum nudum*
    Donald R. Hodel: *Chamaedorea brachypoda, Chamaedorea cataractarum, Chamaedorea metallica, Chamaedorea microspadix*
    Nina Magoun: *Colvillea racemosa*
    Paul R. Weissich: *Clerodendrum quadriloculare, Hibiscus tiliaceus* var. *sterile* and variegated forms

We are, however, most deeply indebted to Dr. George W. Staples III, Botanist, and Clyde T. Imada, Research Assistant, of the Bishop Museum's *In Gardens of Hawai'i* project. We were given access to their many years of research, enabling us to follow the nomenclature of that publication. Our readers will be able to use both publications without the confusion of dissimilar plant names. They must join us in recognizing and thanking those two botanists for their professional support, generosity, and limitless patience.

Mahalo nui loa to you all.

Fred D. Rauch
Paul R. Weissich

Chapter 1
Ground Covers

Ground covers provide a low carpet of dense growth ranging in height from a few inches to several feet. They function to prevent erosion, control weeds, and protect the roots of trees and shrubs from extremes in temperature. They are an important design element in the landscape. Ground covers are commonly used as substitutes for grass, especially in areas where it is difficult to maintain or grow grasses, such as deep shade or steep slopes.

Agapanthus praecox subsp. *orientalis*
AFRICAN LILY
Liliaceae (Lily Family)

An evergreen South African perennial reaching 4 feet high, this slow growing plant is an excellent garden or tub plant for sunny or partially shaded locations. Several cultivars are available, including dwarf forms and those with flowers of dark blue to light blue and white produced in late spring. They are excellent as cut flowers and can be used in lei making. The African lily is heat tolerant and partially salt and drought tolerant.

Aloe vera
BARBADOS ALOE, ALOE VERA, PĀNINI ʻAWA ʻAWA
Aloeaceae (Aloe Family)

Clumping to 2 feet high, this species produces flowering stalks 3 feet high in winter or spring. This plant, probably from the Mediterranean region, is drought, wind, and salt tolerant and does best in a light, well-drained soil in sun but tolerates shade. It is also useful for rock gardens and as a potted specimen. It is used as a home remedy for burns and for skin irritations. Many other aloe species are available. Flowers and entire plants are incorporated into arrangements.

Alternanthera tenella
JOYWEED
Amaranthaceae (Amaranth Family)

This evergreen plant from Mexico to Argentina grows to 12 inches high. It is used for edging or special landscape designs in full sun. Many cultivars are available that are used primarily for their colored foliage, which may be green with yellow, green with red, or green with white variegation.

Aptenia cordifolia hybrid
HEARTS AND FLOWERS
Aizoaceae (Ice Plant Family)

A succulent South African native, this plant spreads by creeping stems that reach 2 feet in length. In full sun in a well-drained soil, it flowers throughout the year and produces a dense mat to 12 inches deep. Cultivars are available with white borders on the leaves and one with white flowers. It is useful as a hanging basket or window box subject.

Arachis pintoi 'Golden Glory'
GOLDEN GLORY, PERENNIAL PEANUT, PINTO PEANUT
Fabaceae (Bean Family)

This South American native forms an excellent cover to 6 inches deep, rooting at nodes as it spreads. It is especially effective as a bank cover, thriving in full sun. It is a rapid grower, will tolerate some drought, but has little salt tolerance. It can also be used in window boxes and other planters.

Asparagus densiflorus 'Myers'
MYERS ASPARAGUS, FOXTAIL ASPARAGUS
Liliaceae (Lily Family)

An evergreen, this South African native forms clumps that reach 2 feet in height. It is also useful as a container plant, for planter boxes, in rock gardens, and will thrive in both sun and shade. It is somewhat drought tolerant, has moderate salt tolerance, and will grow readily in any well-drained soil. Stems are used in arrangements.

Asparagus densiflorus 'Sprengeri'
SPRENGER ASPARAGUS
Liliaceae (Lily Family)

From South Africa, this species forms clumps of arching stems that reach 6 feet in length. It will do well in sun or shade, in most soils, and has good drought and salt tolerance. It also makes a good potted specimen or hanging basket subject. Stems are used in arrangements. There is a compact form with stems to 2 feet long. Eradicate seedlings to prevent this species from becoming invasive.

Asystasia gangetica
ASYSTASIA, COROMANDEL, GANGES VIOLET
Acanthaceae (Acanth Family)

This vigorous, trailing perennial from Africa, India, and Malaysia grows to 2 feet high, producing flowers of light violet, violet, white, or yellow all year. It is especially useful for covering slopes owing to its soil-binding root system. It grows rapidly in most soils, in sun or partially shaded locations, and is highly drought tolerant.

Bacopa monnieri
BACOPA, WATER HYSSOP, ʻAEʻAE
Scrophulariaceae (Snapdragon Family)

Rooting at each node, ʻAeʻae forms a dense mat several inches thick, grows well in moist sand or soil, and prefers full sun. It is native to Hawaiʻi as well as many other tropical regions in coastal areas, where it colonizes in mud and sand flats, marshes, and banks of brackish streams.

Catharanthus roseus
MADAGASCAR PERIWINKLE (P)
Apocynaceae (Dogbane Family)

Reaching 2 feet high, this Madagascan shrub produces an abundant bloom throughout the year. Flowers range in color from white, red, and rose pink to purple. It is best grown in full sun in most garden soils. It is also useful as a border, filler plant, or potted specimen. It displays good wind, drought, and salt tolerance.

Chlorophytum comosum
SPIDER PLANT, AIRPLANE PLANT, BRACKET PLANT
Liliaceae (Lily Family)

This clumping plant from South Africa spreads quickly by its flowering stems, which bend to the ground and take root. It makes an excellent house plant or ground cover in well-drained, shaded locations with organic matter. Cultivars are available with yellow or white stripes along the margins or down the center of the leaves. It is frequently used in hanging baskets and window boxes. Entire plants are used in arrangements.

Coprosma x *kirkii* 'Variegata'
VARIEGATED PROSTRATE COPROSMA
Rubiaceae (Coffee Family)

Growing to 3 feet high, sometimes nearly prostrate, this variegated cultivar from New Zealand is tolerant of a wide range of soils, sea wind, and salt spray. It thrives in full sun or light shade, making a dense ground or bank cover for erosion control.

Cuphea hyssopifolia
FALSE HEATHER, MEXICAN HEATHER
Lythraceae (Crape Myrtle Family)

A compact, shrubby evergreen from Mexico to Panama, this species grows to 2 feet high, bearing flowers much of the year. It is best in sun in most soils. It is also used as a border plant or for containers. Several cultivars are available, including a compact form that reaches 1 foot in height and bears tiny rose-purple blooms; another has white flowers.

Dissotis rotundifolia
DISSOTIS
Melastomaceae (Melastome Family)

Native to tropical West Africa, this rapidly growing species forms a dense mat to 1 foot deep and displays showy blooms much of the year. It grows best in sun or partial shade in a well-drained, moist soil. It is especially suited to small, sheltered, well-watered sites and can be used in hanging baskets and window boxes.

Ground Covers

Elatostema repens
TRAILING WATERMELON BEGONIA, PELLIONIA
Urticaceae (Nettle Family)

This creeping perennial from Vietnam grows to 1 foot high, spreading by means of stems that are up to 2 feet long. It is best in a well-drained, moist soil in partial shade. Widely used as a container or hanging basket plant, it is useful as a colorful ground or bank cover.

Evolvulus glomeratus subsp. *grandiflorus*
BLUE DAZE
Convolvulaceae (Morning Glory Family)

Southern Brazil is the home of this densely growing, spreading shrub that reaches 18 inches in height. Flowering best in full sun, it produces small, bright blooms most of the year. It makes a good bank cover and also finds use as a border plant or in the rockery. It has moderate drought and low salt tolerance.

Farfugium japonicum
LIGULARIA, FARFUGIUM
Asteraceae (Sunflower Family)

A clumping evergreen from Japan, this plant grows to 2 feet high. Flowers are produced on stems 1 to 2 feet long during spring and summer. It thrives in partially shady, moist gardens. It may also be used in borders or planters or as a potted specimen. Several cultivars are available that display variegated, speckled, and crinkly foliage.

Gazania rigens var. *leucolaena*
TRAILING GAZANIA
Asteraceae (Sunflower Family)

This trailing perennial plant reaches 1 foot in height and comes from South Africa. It blooms much of the year, more profusely at cooler elevations. It is at its best in full sun in most well-drained soils. In addition to its ground covering capability, it may be used in hanging baskets or window planters.

Heliotropium anomalum var. *argenteum*
HINAHINA
Boraginaceae (Borage Family)

Native to Hawai'i, this species forms a dense, matlike cover up to 3 inches deep and grows in pure sand or well-drained soil. It does best close to the beach in full sun and will tolerate heat, drought, salt air, and direct salt spray. Flowers and foliage are used in lei making.

Hemerocallis aurantiaca
GOLDEN SUMMER DAYLILY
Liliaceae (Lily Family)

A moderately fast growing species from China, profusely flowering in spring and summer, this clump-forming ground cover reaches 2 feet high. Flowers are carried on stems 1 foot above the foliage. Growing best in high light in a rich garden soil, it is adaptable to a wide range of moisture and soil conditions. It is useful in beds, borders, planters, color accent containers, as filler, and especially for covering slopes. Flowers are used in arrangements. There are innumerable cultivars ranging from 1 to 3 feet in height, with varied flowering periods and flowers in yellows, pinks, and reds. Black-Eyed Stella Daylily (*H.* 'Black-Eyed Stella') is a compact grower reaching up to 1 foot high and flowering much of the year. Uses and cultural requirements are the same. It is also seen in hanging baskets.

Hemigraphis alternata
METAL LEAF, RED IVY
Acanthaceae (Acanth Family)

Rooting as it spreads, this Malaysian ground cover grows to 6 inches high in partial shade in a rich, moist, well-drained soil. It is used to provide color in the landscape. It is a good slope cover and also makes a good hanging basket or window box subject. Cut stems are used in arrangements. Several cultivars are available that display various leaf forms. One, the Waffle Plant (*H. alternata* 'Exotica') from New Guinea, is the same height but is more robust, with crinkled foliage and the same growth requirements and uses.

Impatiens hawkeri
NEW GUINEA IMPATIENS
Balsaminaceae (Impatiens Family)

A varied group of striking cultivars has been developed from this New Guinea species. The cultivars vary from upright to spreading habits, with leaves often variegated with cream or red, and with large flowers in various colors including lavender, purple, pink, orange, red, and white. They are best in a cool situation with strong light and a moist, well-drained soil. They are used as bedding plants, for borders, and on slopes and make excellent potted specimens for strong color accents. Some adapt to full sun.

Impatiens wallerana
IMPATIENS, BUSY LIZZIE, ZANZIBAR BALSAM
Balsaminaceae (Impatiens Family)

A vigorous, erect, succulent plant from Tanzania to Mozambique reaching a height of 3 feet, this old-fashioned favorite flowers almost constantly. Numerous cultivars have been developed, including those with single and double flowers in a range of colors and bicolors from white through pink, salmon, and orange to purple. Dwarf forms are also available. They thrive in light shade in moist soil. They make a colorful cover and are useful as potted plants, in hanging baskets, and in window boxes.

Ipomoea pes-caprae subsp. *brasiliensis*

BEACH MORNING GLORY, PŌHUEHUE

Convolvulaceae (Morning Glory Family)

This indigenous species commonly found along Hawaiian coastal areas is also native throughout the tropics. A vine, it spreads rapidly, rooting at each node, forming a matlike ground cover 1 foot deep. Its best use in the landscape is in beach gardens in full sun. Growing in pure sand or any well-drained soil, it is completely salt and wind tolerant. Hawaiians make a special lei of *Pōhuehue*.

Jaquemontia ovalifolia subsp. *sandwicensis*

PĀʻŪOHIʻIAKA

Convolvulaceae (Morning Glory Family)

This diminutive, rapidly spreading, endemic vine is native to sites close to the shore. Salt, wind, and drought tolerance make this ground cover, 2 inches high, useful in beach gardens. It thrives in sand, soil, or among rocks, rooting at each node. It blooms throughout the year, and flower color varies from sky blue to pale blue to white. Foliage varies from green to silvery gray.

Juniperus procumbens

JAPANESE GARDEN JUNIPER, SHORE JUNIPER

Cupressaceae (Cypress Family)

Moderately slow growing, this evergreen, mounding shrub from Japan grows to 2 feet high, spreading up to 6 feet. It thrives in full sun in most soils but requires good drainage. It has moderate drought and salt tolerance. Several cultivars are available. All are useful covers for wall-top plantings, in the rockery, or as potted specimens. Foliage may be used in arrangements.

Lantana montevidensis

TRAILING LANTANA

Verbenaceae (Verbena Family)

This South American shrub grows to 2 feet high and spreads rapidly to 6 feet in width. Flowers are produced in profusion almost continuously. Growing best in full sun in almost any soil, it is drought and heat tolerant and moderately salt tolerant. In addition to ground covering capability, it is used in planter boxes and in hanging baskets. Other color forms are available.

Lipochaeta integrifolia

NEHE

Asteraceae (Sunflower Family)

One of twenty species endemic to the Hawaiian Islands, this rapidly spreading plant grows to 3 inches high. It forms a dense beach garden ground cover, its stems rooting as they grow. It requires full, hot sun and perfect drainage and will grow in both sand and soil. It is highly salt, wind, and drought tolerant. Flowers are used in leis.

Liriope muscari
LILYTURF
Liliaceae (Lily Family)

This slow growing evergreen plant from Japan and China forms dense clumps to 18 inches high. Performing best in medium light in a fertile, well-drained soil with moisture, it is excellent for small areas, on slopes, in borders, or among rock groupings. It makes an attractive potted specimen and is used in arrangements. Several cultivars with various flower colors and white- or yellow-variegated foliage are available.

Neomarica gracilis
WALKING IRIS
Iridaceae (Iris Family)

A perennial plant from South Africa, this species forms dense clumps to 2 feet high. Flowers, lasting one day, are borne on stems above the clump that gradually arch to the ground, taking root and forming a new plant. It is at its best in medium light in a moist, well-drained soil. It makes a good bank cover for erosion control and can also be used as a potted accent for the shaded lanai.

Ophiopogon jaburan 'Vittatus'
WHITE LILYTURF, JABURAN LILYTURF
Liliaceae (Lily Family)

A dense, clumping evergreen plant from Japan, this species grows to 2 feet high. It performs best in sun or medium light in a fertile, well-drained soil with moisture. It is a moderate grower. Also referred to as White Dragon, this cultivar has white-striped leaves. It is a good cover for banks, color accent beds, and as a potted specimen. Whole plants can be used in arrangements.

Ophiopogon japonicus
MONDO GRASS, DWARF LILYTURF
Liliaceae (Lily Family)

This slow growing perennial plant from eastern Asia forms a dense, grasslike clump to 1 foot high. It does well in partial shade in a fertile, well-drained soil, but is adaptable to many soil conditions. Although not tolerant of heavy foot traffic, it is a visual substitute for grass. It is also useful for borders or edging and for erosion control on banks once established. It has moderate drought and salt tolerance. A slow growing dwarf cultivar, Dwarf Mondo Grass (*O. japonicus* 'Compactus') from Japan, forms a dense matlike cover to 4 inches high. It is an excellent ground cover for sun or light shade and for use between stepping-stones and also makes an attractive potted specimen.

Ground Covers

Peperomia obtusifolia
BABY RUBBER PLANT, PEPPER FACE, AMERICAN RUBBER PLANT
Piperaceae (Pepper Family)

A succulent, tropical American perennial to 18 inches high, this moderately fast growing plant is best in low to medium light and in soils with excellent drainage. It is useful under trees, for borders or edging, hanging baskets, and as a container plant for deck or interior. Several dwarf or variegated forms are available, as well as many other species of peperomia that are also useful in the shaded garden.

Pilea cadierei
ALUMINUM PLANT
Urticaceae (Nettle Family)

An erect perennial from Southeast Asia that grows rapidly to 18 inches high, this species makes an excellent ground cover, bank cover, or planter box and container subject. Plant in medium shade in a moist, well-drained soil. A dwarf cultivar is available.

Pilea depressa
BABY'S TEARS
Urticaceae (Nettle Family)

A low, creeping perennial from the West Indies, this species grows to 4 inches high, rooting where it touches the soil and forming a dense matlike cover. This easily grown plant requires moderate shade and moisture in a well-drained soil. It is a useful container, terrarium, or hanging basket plant and is well used as a ground and bank cover.

Pilea nummulariifolia
CREEPING CHARLIE
Urticaceae (Nettle Family)

A low, creeping perennial growing to 8 inches high, this species from Peru, Panama, and the West Indies roots at each node, forming a thick mat. This rapid grower is best in partial shade with a well-drained, moist soil. It is also useful as a container or hanging basket plant.

Pilea serpyllacea
ARTILLERY PLANT
Urticaceae (Nettle Family)

An almost fernlike succulent perennial from tropical America growing to 10 inches high, this species forms a thick cover with its many spreading branches. It performs best in light shade in a well-drained soil with moisture. It may also be used for edging, borders, or in a hanging basket or container.

Plumbago zeylanica
WILD PLUMBAGO, 'ILIE'E (P)
Plumbaginaceae (Plumbago Family)

An indigenous species found in the Old World tropics and Hawai'i, *'Ilie'e* occurs in dry coastal areas in soil, lava flows, and sand dunes as well as inland in dry forests. It is a dense, spreading, somewhat woody shrub growing to 3 feet high. It is highly drought tolerant, making it a useful ground cover for difficult areas. Minimal watering maintains the foliage in a bright green condition.

Polygonum capitatum
PINK CLOVER, KNOTWEED
Polygonaceae (Buckwheat Family)

This Asian plant is a rugged, trailing evergreen, growing to 6 inches high and spreading to 20 inches. Its small round flower heads are produced much of the year. This fast grower is a good cover for waste areas in sun or shade. It is at its best in cooler areas and makes a good hanging basket and potted specimen.

Portulaca grandiflora
MOSS ROSE, PORTULACA
Portulacaceae (Moss Rose Family)

A succulent plant growing to 6 inches high, this species rapidly spreads to 18 inches across. Flowers are produced much of the year in a variety of colors, from rose and red to white and yellow, bicolors, and even some with stripes. There are double forms. Plant in full sun in any well-drained soil as a colorful cover in the Xeriscape, in hanging baskets, or in low containers. Moss rose is native to northern South America.

Rosmarinus officinalis
CREEPING ROSEMARY, DWARF ROSEMARY
Lamiaceae (Mint Family)

This is a slow growing, trailing form of the common rosemary. It is usually listed as *R. officinalis* 'Prostratus.' It grows to 2 feet high with an 8-foot spread. Flowers are produced much of the year. Foliage is aromatic and used in cooking. Grows best in full sun in most well-drained soils. In addition to ground and bank covering abilities, it may be used to trail over a wall or as a potted specimen or in a planter. It is a good subject for the Xeriscape.

Ruellia makoyana
MONKEY PLANT, TRAILING VELVET PLANT
Acanthaceae (Acanth Family)

A trailing plant from Brazil growing to 2 feet high, this excellent cover plant produces flowers much of the year. It does best in a shaded location in a well-drained, moist soil. It can also be used as a potted specimen or in a hanging basket or planter.

Ruellia squarrosa
FRINGELEAF RUELLIA

Acanthaceae (Acanth Family)

This vigorous tropical American perennial, growing to 18 inches high, may be used in sun or partial shade in many soil types, with moisture. The binding root system makes it especially useful as a cover for banks and slopes.

Sansevieria trifasciata 'Hahnii'
BIRD'S NEST SANSEVIERIA

Agavaceae (Agave Family)

This low growing variety of the Gold-Banded Sansevieria forms a rosette of leaves up to 6 inches high. This robust, freely suckering plant can be used in sun or shaded locations under various soil conditions. Bird's Nest Sansevieria will tolerate neglect, is suitable as a container plant for the interior landscape, and may be used in arrangements.

Sida fallax
'ILIMA PAPA, KŪ KAHAKAI

Malvaceae (Hibiscus Family)

This is the creeping form of the native *'ilima*, forming a spreading cover to 3 inches high and 2 feet wide. It is highly salt and drought tolerant and does best in full sun in well-drained soils. It is at its best in the beach garden but will grow in a variety of sites. The flowers appear most of the year and are prized for lei making.

Tradescantia pallida 'Purpurea'
PURPLE HEART, PURPLE TRADESCANTIA

Commelinaceae (Dayflower Family)

A spreading perennial from Mexico, reaching 18 inches high, this species is excellent in the Xeriscape as a cover and can be used as a container plant or in a hanging basket on the lanai or well-lighted interior. It does best in a light, well-drained soil with moderate moisture and full sun or light shade for best color.

Tradescantia spathacea
OYSTER PLANT, MOSES IN THE CRADLE

Commelinaceae (Dayflower Family)

A fast growing tropical American succulent to 2 feet high, this colorful species may be used in sun or shade in most soils but is best in rich, moist soil. It makes an excellent ground cover plant, border or edging plant, and potted specimen. The dwarf form, sometimes listed as *T. spathacea* 'Dwarf,' is more frequently used in the landscape. There are also forms with solid green and variegated foliage.

Tradescantia zebrina

WANDERING JEW, HONOHONO

Commelinaceae (Dayflower Family)

A fleshy, trailing Mexican evergreen perennial rooting at the nodes, this species forms a dense ground cover to 1 foot deep. It is best in shade with moisture. Many cultivars are available, with variously colored leaves. Wandering Jew and its relatives provide a dense erosion-controlling ground cover. They also can be used to make handsome hanging baskets or in window boxes.

Verbena x *hybrida*

PERUVIAN VERBENA, COMMON VERBENA, GARDEN VERBENA

Verbenaceae (Verbena Family)

This fast growing tropical American perennial, forming a dense mat to 6 inches deep, produces spreading stems rooting at the nodes. Flowers in red, scarlet, blue, or purple appear most of the year. It requires sun, heat, and well-drained soils for best results and is useful for dry banks, wall tops, hanging baskets, and rock crevices.

Vitex rotundifolia

BEACH VITEX, KOLOKOLO KAHAKAI, PŌHINAHINA

Verbenaceae (Verbena Family)

Native to shorelines of Hawai'i and throughout the Pacific and Indian Oceans, this highly salt, wind, and drought tolerant plant is excellent for beach gardens as well as those inland. Spreading rapidly, it may reach 4 feet high under cultivation, usually 2 feet at the beach. It roots as it spreads, making a dense cover when grown in full sun. The foliage is pleasantly aromatic. Flowers, which appear throughout the year, are used in lei making.

Wedelia trilobata

WEDELIA (S)

Asteraceae (Sunflower Family)

A strong growing, spreading South American plant, this species forms dense mats up to 2 feet deep or more. Bright blossoms are produced much of the year. Best in full sun, it may also be grown in light shade in most soils, with ample water and fertilizer. It is excellent for erosion control on banks or slopes, in planters, or in hanging baskets. It is highly salt tolerant.

Zephyranthes candida

ZEPHYR LILY, WHITE RAIN LILY (P)

Liliaceae (Lily Family)

A slow growing, bulbous South American plant, this species forms clumps to 12 inches high, bearing a profusion of pure white flowers in summer and fall. It performs best in light shade, in a well-drained, moist soil. There are other species available with yellow, rose, and pink flowers. It can be grown as a potted specimen or accent among shorter ground covers or in the rockery.

Chapter 2
Small Shrubs

This grouping of plants includes both woody shrubs and herbaceous plants in the 2- to 6-foot height range.

Achyranthes splendens
Hawaiian Achyranthes, 'Ewa Hinahina
Amaranthaceae (Amaranth Family)

This endemic native species, growing to 6 feet high, is drought and wind tolerant and shows moderate tolerance to salt air. Achyranthes makes an excellent accent in the Xeriscape or can be used as a low hedge or border. It must be grown in full sun, given a well-drained soil, and watered only lightly for best results. It is especially beautiful with night lighting.

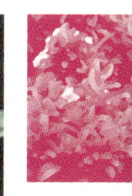

Aglaonema commutatum
Silver Evergreen (P) (S)
Araceae (Aroid Family)

The Philippine Islands and northern Celebes are home to this clumping plant that grows to 3 feet high. It thrives in shaded areas in a fertile, well-drained soil. It shows moderate salt and drought tolerance and is excellent in containers for interior landscaping with moderate light. It can be used as a border or ground cover. There are a number of cultivars with attractive foliage patterns. *A. commutatum* 'Emerald Beauty' displays particularly rich banding, while *A. commutatum tricolor* 'Harlequin' is prized for its patterned silver foliage. *A.* x 'Silver Queen,' a complex hybrid partially parented by *A. commutatum*, is favored for use in deep shade because of its silvery leaves. All are popular for their ease of growth and design functions.

'Emerald Beauty'

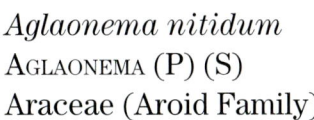
tricolor 'Harlequin'   'Silver Queen'

Aglaonema nitidum
Aglaonema (P) (S)
Araceae (Aroid Family)

An excellent choice for deeply shaded, moist places, this Malaysian Aglaonema tolerates interior conditions as well. Rich, well-drained soil will produce moderate growth and husky clumps to 3 feet high. Use it as a potted specimen, ground cover, bank cover, or for a broad border. Popular cultivars include *A. nitidum* 'Ernesto's Favorite,' with striking, roughly striped leaves; *A.* x 'José Rizal,' with unique speckled foliage; and *A.* x 'Los Baños,' with broad-banded silver leaves. The latter two are complex hybrids involving, in part, *A. nitidum*. All are valued for their ease of growth and usefulness.

'Ernesto's Favorite'

'José Rizal'   'Los Baños'

Alocasia cucullata
CHINESE TARO (S)
Araceae (Aroid Family)

This 2-foot high clumping plant from Ceylon and northeast India to Myanmar is best in filtered light in a moist soil with excellent drainage and wind protection. It is not salt or drought tolerant. It is good used in mass, as a ground cover planting, and makes an excellent container plant for a shaded deck or patio.

Alocasia sanderiana
KRIS PLANT (S)
Araceae (Aroid Family)

A highly ornamental species from the Philippines, this plant grows to 2 feet high. It prefers a moist, rich, well-drained soil in a shady situation protected from wind. Use it as an accent—both in the garden and containerized for the shaded lanai—or for a ground cover or border.

Anthurium hookeri
BIRD'S-NEST ANTHURIUM (S)
Araceae (Aroid Family)

One of several excellent landscape species displaying a "nest" of large leaves, this tropical American grows well in rich, moist soils and as an epiphyte if well watered. Leaves may reach 8 feet long under ideal conditions of light shade, heavy watering, and regular feeding. Use it as an accent in the wind-sheltered shade garden or as a tubbed specimen for the lanai.

Artemisia mauiensis
'AHINAHINA, HINAHINA
Asteraceae (Sunflower Family)

Although endemic to higher elevations on the slopes of Haleakalā, Maui, this 1-foot high, bushy native grows well at low elevations. It requires full sun, excellent drainage, and careful watering to avoid soggy soil. It is used as an accent or ground cover, is suited to the Xeriscape, and makes a good potted specimen.

Bougainvillea 'Temple Fire'
TEMPLE FIRE BOUGAINVILLEA (T)
Nyctaginaceae (Four O'Clock Family)

A shrubby cultivar to 4 feet high and spreading to 6 feet, this dwarf is best grown in full sun in most soils. It has moderate salt and drought tolerance and good tolerance to wind. It may be used as a color accent, ground cover, or potted specimen. A similar dwarf cultivar is *Bougainvillea* 'Menehune,' with pale purple bracts.

Carissa macrocarpa 'Prostrata'
PROSTRATE NATAL PLUM (T)
Apocynaceae (Dogbane Family)

A vigorous, evergreen spreading ground cover to 2 feet high, this moderately drought tolerant cultivar grows well in full sun in a variety of soils. Prune out occasional vertical growth. It has good tolerance to heat, wind, and salt, and makes a good bank cover or container specimen.

Crassula ovata
LARGE JADE PLANT, JADE PLANT, JADE TREE
Crassulaceae (Stonecrop Family)

Growing 12 feet high in its native South Africa, this densely bushy succulent is usually seen as a potted plant or as a larger specimen, to 6 feet high, in the garden. It tolerates heat, drought, wind, and salt air. A number of forms are available, including those with a dwarf habit, blue-green or reddish color, and slender foliage. It makes a useful rockery subject or can be employed as a low hedge or border in the Xeriscape garden. Leaves have been made into leis.

Cuphea ignea
CIGAR FLOWER, FIRECRACKER PLANT, PUA KĪKĀ
Lythraceae (Crape Myrtle Family)

This is a slow growing, evergreen spreading species from Mexico reaching 3 feet in height. It does best in sun in a good soil with moisture but will take most soils and light shade. It has moderate drought and wind tolerance but poor salt tolerance. It is useful as a bank cover, formal border, bedding, and container plant and the flowers are prized for lei making. Several color forms are available.

Cyperus involucratus
UMBRELLA PLANT
Cyperaceae (Sedge Family)

Growing well in full sun or shade in very moist soils or in shallow water, this perennial Madagascan native reaches 3 feet in height. It is a good house plant and an effective accent near pools, in pots (in or out of water), or in planters. This plant forms a clump in the garden that slowly spreads and benefits from cutting back to the ground periodically.

Euphorbia milii
CROWN OF THORNS, CHRIST'S THORN (P) (T)
Euphorbiaceae (Euphorbia Family)

This Madagascan native grows to 5 feet high with equal spread and flowers best when grown in full sun in a light, well-drained soil. It has good drought, salt, and wind tolerance. It may be used in planters or containers, as a foundation plant, low hedge, or bank cover. Both giant and dwarf cultivars are available, with a range of sizes and bract colors from deep red, orange, and coral to pink.

Small Shrubs

Gardenia augusta 'Radicans'
TRAILING GARDENIA, PROSTRATE GARDENIA (P)
Rubiaceae (Coffee Family)

This dense evergreen cultivar grows to 2 feet high with a 3-foot spread. Flowers are fragrant. It performs best in partial shade in acid, well-drained soils. It displays moderate drought and wind tolerance but poor salt tolerance. It makes an excellent container or ground cover plant and fares better in cool valley or hillside areas in Hawai'i.

Heliconia psittacorum
PARROT'S HELICONIA
Heliconiaceae (Heliconia Family)

A spreading South American plant to 4 feet high, this species may be used as a bedding plant or in planters or large containers. It grows in sun or partial shade in most acid soils and has moderate drought, wind, and salt tolerance. The planting location should be selected with care, as rhizomes are invasive and root barriers are recommended. Several newer color forms are available. All are excellent cut flowers.

Heliconia stricta 'Dwarf Jamaican'
DWARF JAMAICAN HELICONIA
Heliconiaceae (Heliconia Family)

This wide spreading northern South American cultivar develops dense clumps 2 feet high, bearing its flowers below the foliage most of the year. It does best in a rich, well-drained soil with moisture and protection from the wind. Use it for a ground or bank cover in the lightly shaded garden. It makes a good container plant and is popular for arrangements.

Iresine herbstii
BLOOD LEAF, BEEFSTEAK PLANT
Amaranthaceae (Amaranth Family)

This tropical South American plant grows to 5 feet high. It does best in high light with a moist, well-drained soil. It is a useful plant for containers, as a bedding plant, or for borders or a low hedge. Many cultivars are available with colorful foliage showing red stems and petioles and green or greenish red leaves with yellow veins.

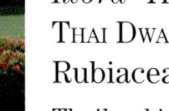

Ixora 'Thai Dwarf'
THAI DWARF IXORA
Rubiaceae (Coffee Family)

Thailand is the source of a great variety of new hybrid ixoras. Among these is a series of dwarf plants growing to 4 feet high and providing a range of brilliantly colored small flowers in red, orange, gold, pink, and yellow borne in abundant, dense clusters throughout the year. They prefer a rich, well-drained soil with regular watering and fertilizing. Full sun produces maximum flowering. Use this plant as a colorful, shrubby ground cover or as a border or facer for larger plants. It performs well as a potted or tubbed specimen and can used in arrangements.

Malpighia coccigera
SINGAPORE HOLLY
Malpighiaceae (Malpighia Family)

Growing to 6 feet high, this West Indian species is useful as a low border or hedge, in planters, or trained in a container; it takes well to severe clipping. It is best in partial shade in a rich, moist, sandy soil with organic matter. It has poor salt tolerance.

Osmoxylon lineare
OSMOXYLON
Araliaceae (Panax Family)

An excellent foliage accent from the Philippines, Osmoxylon makes a unique facer or border plant, reaching 6 feet in height, and it does well in a container. It does best in a rich, well-drained, moist soil in light shade. It is not drought or salt tolerant but resists moderate winds.

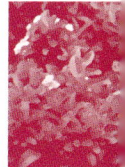

Otocanthus coeruleus
AMAZON BLUE
Scrophulariaceae (Snapdragon Family)

This Brazilian species grows to 6 feet high, but it benefits from heavy pruning, which prevents legginess and serves to force heavier flowering. The flowers are produced all year and are excellent for arranging. Grow it in full sun in a good, moist, acid, well-drained soil. It has moderate drought tolerance but no salt tolerance. Use it as a bedding plant or in a mixed color border with other species used for cutting. It accommodates to container growing.

Pelargonium x *hortorum*
GARDEN GERANIUM, COMMON GERANIUM, LANIUMA
Geraniaceae (Geranium Family)

Originating in South Africa, these hybrids grow to 3 feet high and are used for bedding plants, for planters, or as container specimens. The highly variable leaves may be scalloped, toothed, or variegated, while flower colors range from white through pink, coral, and peach to red. Bicolors and both single and double flowers are available. They are best in sun in a rich, well-drained soil with regular watering. Geraniums tolerate heat and poor, alkaline soil but are not tolerant of salt. Foliage and flowers, which appear all year, are used in arrangements.

Pentas lanceolatus
PENTAS
Rubiaceae (Coffee Family)

Although reaching 6 feet in height, Pentas profits from heavy pruning to prevent legginess and promote maximum flowering. The flowers, varying in color from white through light and dark pink, lavender, and red, appear throughout the year and are good for arrangements. Grow Pentas in the color border with other cutting materials or as a color facer for out-of-season flowering shrubs. It grows best in full sun in a rich, moist, well-drained soil. Its native home is tropical Africa.

Small Shrubs

Pittosporum tobira 'Wheeler's Dwarf'
WHEELER'S DWARF PITTOSPORUM
Pittosporaceae (Pittosporum Family)

Densely mounding to 3 feet high, this dwarf form makes an excellent ground cover, foreground or low boundary planting, or container plant. May be grown in sun or light shade in most soils with good drainage. It has good drought, salt, and wind tolerance.

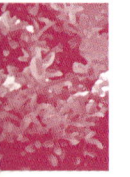

Sansevieria trifasciata
BOWSTRING HEMP, SNAKE PLANT, MOTHER-IN-LAW'S TONGUE
Agavaceae (Agave Family)

This tough plant from Africa grows to 4 feet high. It can be grown in high or low light and tolerates a variety of soils, salt air, drought, and saline soils. It is used as an accent plant and is excellent as a container plant for a sunny lanai or as a house plant. The Gold-Banded Sansevieria cultivar, (*S. trifasciata* 'Laurentii'), has foliage banded in yellow. Flowers are white and used in arrangements, as are leaves and entire plants.

Senecio cineraria
DUSTY MILLER
Asteraceae (Sunflower Family)

From the Mediterranean region and growing to 3 feet high, this plant is best in sun in a well-drained soil. It has good drought, wind, and heat tolerance, as well as moderate salt tolerance, faring better in cooler situations. It is an excellent container plant and also useful for edging, mass planting, bedding, or large ground cover plantings.

Serissa japonica
SERISSA
Rubiaceae (Coffee Family)

This slow growing species from Southeast Asia reaches 3 feet in height, producing small white flowers from spring to fall. It is used as a specimen, in a low hedge, for mass planting, or as a bonsai subject. It will grow in sun or partial shade in most well-drained soils and has good salt, drought, and wind tolerance.

'McCoy'

Spathiphyllum spp.
SPATHIPHYLLUM
Araceae (Aroid Family)

A number of species of spathiphyllum, or spathe flower, are available for garden use. Most species are from the New World tropics, several from Malaysia. They vary in size from dwarfs only a few inches in height with pale green, shaded foliage to 8-foot tall species with leaves ranging from light green to dark green. Flowers are white or greenish white, some fragrant. All require shade, a rich, loamy soil, protection from wind, and ample watering. Among the most available are the following:

Spathiphyllum cannifolium bears broad, straight, dark green leaves forming a dense head to 4 feet high, with white to greenish white fragrant flowers;

S. 'McCoy,' a locally developed cultivar, growing to 4 feet high, displays large, dark green, curving leaves and large, white to greenish white blooms held above the foliage;

S. 'Sensation,' another large cultivar, is especially valued for its low light tolerance and use in interiorscapes;

S. wallisii 'Clevelandii,' a hybrid and locally the most commonly seen of the group, grows to 2 feet high, with pure white flowers found plentifully throughout the year and leaves that are shiny, dark green, and arching downward.

All make excellent potted or tubbed plants for the shaded lanai or terrace. They are all useful in a tropical border, massed into an attractive ground cover or bank cover and the larger species used as accent and background planting for smaller species. All are used in arrangements.

'Sensation'

wallisii 'Clevelandii'

Turnera ulmifolia
YELLOW ALDER, SUNDROPS
Turneraceae (Turnera Family)

This tropical American species grows to 3 feet high, producing its fragrant flowers much of the year. It may be used as a large evergreen ground cover, as well as a border or filler plant for sun or partial shade in the garden. It prefers a well-drained soil and has good drought tolerance.

Wikstroemia uva-ursi
'ĀKIA (P)
Thymelaeaceae ('Ākia Family)

An endemic Hawaiian species, '*Ākia* forms a dense, shrubby, spreading ground cover to 3 feet high with equal spread. '*Ākia* is wind, drought, and salt tolerant and prefers full, hot sun. It responds to pruning and shaping and makes a good potted specimen, as well as a good bank or ground cover. It will grow in a variety of well-drained soils. The foliage, flower clusters, and small, round, red fruit can be used in leis.

Chapter 3
Medium Shrubs

This grouping of plants includes both woody and herbaceous plants generally in the 6- to 10-foot height range. Under ideal conditions, however, some may exceed those heights.

Abelia × *grandiflora*
GLOSSY ABELIA
Caprifoliaceae (Honeysuckle Family)

A Chinese species reaching 10 feet in height, this evergreen is used as an informal hedge, screening, or background plant, as well as a foundation planting. White fragrant flowers appear in summer and early fall and may be cut for arrangements. It is best in high light in a well-drained soil, but is tolerant of light shade. It has moderate drought tolerance but no salt tolerance.

Acalypha godseffiana 'Heterophylla'
ACALYPHA, THREE-SEEDED MERCURY
Euphorbiaceae (Euphorbia Family)

A favored medium hedge that grows to 8 feet high, this plant from New Guinea displays two foliage color forms: green with yellow leaf margins and green with red leaf margins. Its leaves are elongated, slender, and shaggy, with somewhat of a weeping habit. It thrives best in full sun in rich, moist, well-drained soil. Moderately rapid growing, this shrub makes a dense hedge or privacy screen and is also good as a container specimen.

Acalypha hispida
CHENILLE PLANT, RED HOT POKER (P)
Euphorbiaceae (Euphorbia Family)

A dense evergreen from Indonesia growing to 8 feet high, the Chenille Plant is best planted in good, rich, moist soils, in sunny or lightly shaded places. Its colorful inflorescences are displayed much of the year. It benefits from protection from wind and has little salt tolerance. It makes an interesting accent or specimen plant and may also be used for background or enclosures or as a container plant.

Adenium obesum
ADENIUM, DESERT ROSE, MOCK AZALEA (P) (S)
Apocynaceae (Dogbane Family)

Dry areas of East Africa and adjacent Arabia are home to several species of adenium, which are useful in the landscape for color and form accents in the Xeriscape, the rockery, or in containers. Desert Rose grows slowly to 10 feet high. It must have a well-drained, light soil for best results and moderate watering to insure year-round flowering. It is heat, drought, and salt-wind tolerant. Flower color ranges from dark pink and white to light pink and white to pure white.

Agave attenuata
Dragon Tree Agave, Swan's Neck Agave
Agavaceae (Agave Family)

This slow growing evergreen plant from Mexico forms a rosette of leaves 1.5 to 2.5 feet long on a trunk to 5 feet high. The entire plant, with its offsets, is used in arrangements. The Dragon Tree Agave does best in a sandy soil with good drainage and moisture, but it is tolerant of poor soils, drought, salt, wind, and even fire. It is a good accent near pools or in a beach landscape, either singly, in masses, or in containers. This is one of many agave species available to the gardener.

Allamanda blanchetii
Purple Allamanda (P)
Apocynaceae (Dogbane Family)

A sprawling evergreen Brazilian species, this plant grows to 6 feet high in full sun and most soils and produces flowers in summer and fall. It may be used as an informal hedge or screen plant or as a large bank or ground cover. It has moderate drought and wind tolerance but is not tolerant of salts.

Allamanda schotti
Bush Allamanda (P)
Apocynaceae (Dogbane Family)

A slow growing evergreen Brazilian shrub reaching 5 feet in height, this allamanda finds good landscape use as an informal hedge, bank cover, or filler plant. It flowers best in a fertile, well-drained soil in full sun. It has moderate drought and salt tolerance and good tolerance to wind.

Alpinia purpurata
Red Ginger, 'Awapuhi 'Ula'ula
Zingiberaceae (Ginger Family)

This evergreen clumping plant from Melanesia attains a height of 10 feet when grown in sun or light shade in a rich, moist soil with protection from wind. Its flowers are excellent for arrangements. Several cultivars are available, displaying light pink to dark pink and double red inflorescences.

Anisacanthus thurberi
Desert Honeysuckle
Acanthaceae (Acanth Family)

This evergreen shrub from the southern United States grows to 5 feet high, displaying its flowers much of the year. It may be used for mass or filler planting, informal hedges, or as a tall ground cover. It is best in sun in a well-drained soil; it has good drought and wind tolerance, as well as moderate salt tolerance.

Ardisia crenata
HILO HOLLY, HEN'S EYES
Myrsinaceae (Marlberry Family)

This slow growing evergreen shrub from northeast India to Japan may reach 6 feet in height. Its bright red berries, produced in the winter, are used in arrangements. It is best grown in a moist, shaded garden as a specimen plant, in masses, as a low hedge, or as a container plant. It is also used in rock gardens and in Oriental gardens.

Coprosma repens 'Picturata'
VARIEGATED MIRROR PLANT
Rubiaceae (Coffee Family)

A fast growing evergreen shrub, mounding to 10 feet in height, this species is used as a screen, mass planting, specimen, or hedge. This New Zealand native is best in full sun in cool elevations. It has good drought and wind tolerance and some salt tolerance.

Crinum asiaticum
SPIDER LILY, GIANT LILY, POISON BULB (P)
Liliaceae (Lily Family)

This evergreen plant from Southeast Asia forms broad clumps to 5 feet high used for mass plantings, borders, and for container growing. It produces large clusters of fragrant flowers most of the year. Foliage and fruit may be tinged with purple. It is at its best in sun in a well-drained soil but is tolerant of poor soils, salt, drought, and light shading. Flowers, foliage, and fruits are used in arrangements.

Dichorisandra thyrsiflora
BLUE GINGER
Commelinaceae (Dayflower Family)

This clumping Brazilian evergreen plant produces its striking blue flowers throughout the year on 8-foot stems. It is at its best in a moist, lightly shaded garden in a soil rich in organic matter. Blue Ginger is useful as a color specimen, a hedge, or in mass plantings.

Dieffenbachia amoena
DUMB CANE, DUMB PLANT (S)+
Araceae (Aroid Family)

When provided with light shade, deep, well-drained, moist soil, and protection from wind, this striking Dumb Cane will grow over 6 feet high. It is a bold accent or background border plant. It makes an excellent tubbed specimen and will tolerate interior use in an area with strong light. A popular cultivar is *D. amoena* 'Tropic White.'

'Rudolph Roehrs'

Dieffenbachia maculata
Dumb Cane, Dumb Plant (S)+
Araceae (Aroid Family)

Coming from rain forest areas of tropical America, this species requires moisture and a rich, well-drained soil in a partially shaded location in a protected place. Forming thick clumps to 6 feet high, Dumb Cane makes an excellent accent or tall border plant and can be containerized for use on the lanai or interior. There are a number of handsome cultivars, among which are *D. maculata* 'Rudolph Roehrs,' 'Superba,' and 'Uleryii.'

'Superba' 'Uleryii'

Eranthemum pulchellum
Blue Sage, Blue Eranthemum
Acanthaceae (Acanth Family)

Up to 6 feet high, this evergreen from Southeast Asia shows its bright blue flowers in winter and early spring. It is useful in mass plantings, as a hedge, or as a low screen and makes a good container plant. It is best in light shade in a well-drained, moist soil. It benefits from wind and salt protection.

Euryops pectinatus x *E. abrotanifolius*
Euryops
Asteraceae (Sunflower Family)

A fast growing South African evergreen shrub reaching 6 feet in height, this plant produces its bright flowers much of the year. It thrives in full sun in any good soil, requires excellent drainage, and has good salt and drought tolerance. It is a colorful filler plant, background plant, or low screen.

Ficus deltoidea
Mistletoe Fig
Moraceae (Mulberry Family)

This very slow growing evergreen shrub from Malaysia eventually reaches a height of 10 feet. Mistletoe Fig does best in partial shade and tolerates many soil conditions. It is commonly used as a container plant for the deck or interior and is useful as a specimen, hedge, or screen. It is moderately drought and salt tolerant.

Galphimia gracilis
Galphimia, Rain of Gold, Spray of Gold
Malpighiaceae (Malpighia Family)

A Mexican and Central American native, this species produces its fragrant yellow flowers in abundance throughout the year on an evergreen plant reaching 9 feet in height. It responds best when planted in full sun in a well-drained soil. It tolerates heat and low fertility and has moderate drought resistance but poor salt tolerance. It is a versatile landscape plant for foundation, background, or enclosure plantings.

Gardenia augusta
Common Gardenia, Cape Gardenia, Kiele (P)
Rubiaceae (Coffee Family)

This variable evergreen Chinese shrub, reaching 8 feet in height, produces highly fragrant blooms in late spring. Rich, acid, well-drained soil high in organic matter offers best results. It flourishes in sun in cooler areas, but should have filtered light in hot locations. A favorite specimen plant for the garden, it may be used in containers. It is also useful as a background, hedge, or foundation plant. The flowers are used in arrangements and leis.

Gossypium tomentosum
Hawaiian Cotton, Ma'o, Huluhulu
Malvaceae (Hibiscus Family)

This is a native Hawaiian evergreen species that grows to 6 feet high and produces bright yellow blooms much of the year. It is adapted to hot, dry coastal areas and is best in full sun in a well-drained soil. Uses include as a specimen, a low hedge, large ground cover, or part of a shrub composition. The flowers are used in lei making.

Graptophyllum pictum
Caricature Plant
Acanthaceae (Acanth Family)

Producing small clusters of crimson purple flowers much of the year on a 10-foot tall, compact plant, this easily grown shrub is probably a native of New Guinea. It may be used in sun or shade for accent, in groupings, or as a hedge or screening plant. It does best in a fertile, moist soil but tolerates a range of conditions. It is not salt tolerant. Another common form has purple foliage with irregular pink variegation in the leaf center. The cultivar Eldorado (*G. pictum* 'Eldorado') displays dark green leaves with an irregular, yellow margin that is bright yellow in high light but yellow green in low light. It is an excellent container plant, color accent, hedge, or screen plant.

Heliconia rostrata
Parrot's-Beak Heliconia
Heliconiaceae (Heliconia Family)

Originating in Amazonian Peru and Ecuador, this species flowers all year and is an excellent cut flower. In good moist soil and full sun or light shade, it rapidly attains 10 feet in height. Protection from wind and salt is beneficial. It makes a strong color statement, while its foliage forms an attractive background for smaller tropicals.

Ixora spp.
Ixora
Rubiaceae (Coffee Family)

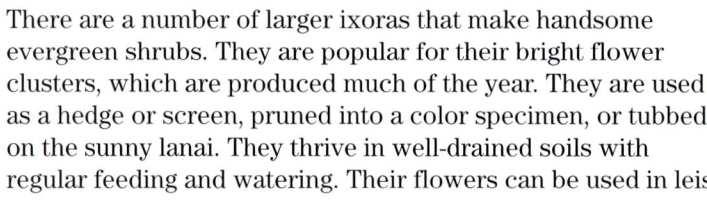

There are a number of larger ixoras that make handsome evergreen shrubs. They are popular for their bright flower clusters, which are produced much of the year. They are used as a hedge or screen, pruned into a color specimen, or tubbed on the sunny lanai. They thrive in well-drained soils with regular feeding and watering. Their flowers can be used in leis.

Ixora casei 'Super King'

Ixora chinensis 'Nora Grant'

Super King Ixora (*Ixora casei* 'Super King') is native to the Caroline Islands of the western Pacific. This shrub grows to 12 feet high and bears very large, bright red flower clusters up to 6 inches across.

Nora Grant Ixora (*Ixora chinensis* 'Nora Grant'), from south China into the Malay Peninsula, is a shrub to 12 feet high.

Red Ixora (*Ixora coccinea*) grows to 10 feet high and displays flowers in the red and red orange range and into pink, yellow, and white. Native to Sri Lanka and southern India, it is seen in both informal and formal hedges (clipping stimulates new growth and flower production), as espaliers, and as potted specimens.

Ixora coccinea

Jasminum sambac
Arabian Jasmine, Pīkake (P) (A)
Oleaceae (Olive Family)

This sprawling woody shrub from India, reaching a height of 5 feet, produces clusters of fragrant flowers throughout the warm months of the year. They are prized by lei makers. *Pīkake* is best grown in warm, dry locations in full sun and a well-drained soil. It is excellent for mass, filler, and foundation plantings. Large-flowered and double forms are available.

Juniperus chinensis 'Pfitzeriana'
Pfitzer Juniper
Cupressaceae (Cypress Family)

This coniferous plant, possibly from China, rapidly reaches 10 feet in height and spreads twice that much. It is well used as a specimen, in mass plantings, in the rock garden, and for screens where space permits. Pfitzer Juniper is best in sun and tolerates most well-drained soils. It has good salt and wind tolerance and moderate drought tolerance. The foliage can be used in arrangements.

Justicia brandeegeana
SHRIMP PLANT
Acanthaceae (Acanth Family)

A native of Mexico, this evergreen species grows to 6 feet high and displays flowers throughout the year. Adaptable, it may be used in sun or partial shade but is best in a rich, well-drained soil. It is an excellent plant for mass plantings in front of taller shrubs or as a colorful border plant. It also makes a good tub or container plant for the interiorscape or landscape. It is partially drought tolerant. There is also a yellow-flowered form.

Justicia carnea
KING'S CROWN, JACOBINIA, FLAMINGO FLOWER
Acanthaceae (Acanth Family)

An evergreen from northern South America growing to 5 feet high, this adaptable plant grows best in rich, moist, well-drained soil, in sun or part shade where its bright flowers are seen most of the year. It is used as a garden or container specimen plant, in mass color plantings, or as a low hedge.

Kalanchoe beharensis
VELVET LEAF, FELT BUSH
Crassulaceae (Stonecrop Family)

A succulent species from Madagascar that grows 10 feet high, this tough plant is best grown in a well-drained soil in full sun, but it will tolerate light shade. It has good drought, heat, wind, and salt tolerance and makes a striking specimen in the rock garden or as a potted accent on a sunny lanai or terrace.

Lantana camara
LANTANA, BUSH LANTANA, LĀKANA (P) (T)
Verbenaceae (Verbena Family)

This fast growing evergreen shrub from the American tropics grows to 6 feet high, producing its flowers most of the year. It is best in full sun in most well-drained soils and has good drought, salt, heat, and wind tolerance. It makes a useful color mass, specimen, or foundation plant but may also be used in containers and hanging baskets. Cultivars are available with flower colors of white, yellow, orange, and red, as well as dwarf forms.

Medium Shrubs

Leucophyllum frutescens

TEXAS RANGER

Scrophulariaceae (Snapdragon Family)

Compact, slow growing, and evergreen, this species from Texas and Mexico grows to 12 feet high, preferring full sun in a well-drained soil. It tolerates heat, wind, and drought. Use as a shrub mass, clipped hedge, informal screen, or background plant. Several selections are available with more silvery foliage and various flower colors. It flowers abundantly during the summer.

Manihot esculenta 'Variegata'

VARIEGATED CASSAVA (P)

Euphorbiaceae (Euphorbia Family)

A fast growing Brazilian shrub reaching 8 feet in height, this evergreen can be used as a color accent or specimen, in mass plantings, or as a container plant. This relatively short-lived plant may be used in sun or light shade and is best in a rich, moist, well-drained soil.

Medinilla magnifica

MEDINILLA, MAGNIFICENT MEDINILLA

Melastomaceae (Melastome Family)

This evergreen plant from the Philippines reaches 10 feet in height and flowers during spring and summer. Medinilla requires a cool, moist location in filtered light and protection from wind. This slow growing plant makes a spectacular garden specimen, container plant, or massed planting.

Mussaenda frondosa

MUSSAENDA

Rubiaceae (Coffee Family)

This evergreen plant from India and the Andaman Islands grows to 9 feet high and makes an excellent specimen plant, screen, or mass planting. It thrives in light shade but will tolerate sun, and it likes moisture and protection from strong wind. It is not salt tolerant. It flowers from spring through fall.

Nandina domestica

HEAVENLY BAMBOO, SACRED BAMBOO, NANTEN (P)

Berberidaceae (Barberry Family)

A slender, clumping evergreen reaching 8 feet in height and originating in China and Japan, this shrub grows well in sun or shade and prefers a moist, rich, well-drained soil. Flowering is in early spring and summer followed by bright red fruit. It makes an excellent vertical accent and potted specimen. All parts of the plant are used in arrangements. Several cultivars are available, including dwarf forms.

Nototrichium sandwicense

KULUĪ

Amaranthaceae (Amaranth Family)

This highly variable Hawaiian species, reaching 10 feet in height, thrives in full sun in a light, well-drained soil. It has good drought, wind, and salt tolerance. Use *Kuluī* as an accent, hedge, screen, windbreak, or in a container for the sunny lanai. It is highly useful in the night garden.

Ochna thomasiana

MICKEY MOUSE PLANT

Ochnaceae (Mickey Mouse Plant Family)

A slow growing evergreen reaching 10 feet in height, this useful tropical East African plant flowers much of the year, most heavily in summer. The shape of the fruit gives rise to its common name. It flourishes in light shade in a rich, moist, well-drained soil. It may be used as a specimen, hedge, screen, or container plant.

Odontonema cuspidatum

RED JUSTICIA, ODONTONEMA, FIRE SPIKE

Acanthaceae (Acanth Family)

An evergreen from Central America, this shrub grows to 10 feet high and prefers sun or light shade in a rich, moist, well-drained soil. It produces flowers most of the year and may be used as a specimen, in foundation or mass plantings, or as a formal or informal hedge or screen.

Osteomeles anthyllidifolia

ʻULEI

Rosaceae (Rose Family)

This Hawaiian native species thrives in both moist and dry areas, is wind, salt, and drought tolerant, and reaches 10 feet in height. It is more compact in full sun but will tolerate light shade. Generally rambling in nature, it makes a tough ground cover, can be pruned into a dense hedge or screen, and is a good potted specimen. Foliage, flowers, and fruit are used by lei makers.

Pachystachys lutea

LOLLIPOP PLANT, YELLOW LOLLIPOP

Acanthaceae (Acanth Family)

This evergreen shrub from Peru and Ecuador grows to 8 feet high and produces flowers much of the year. It performs best in good, well-drained soil that should not be permitted to dry out. It is excellent for a dash of color in the garden, as a low hedge, a tubbed specimen, or in a shrub border. Plant in sun or light shade.

Medium Shrubs

Pittosporum tobira 'Variegata'
VARIEGATED JAPANESE PITTOSPORUM
Pittosporaceae (Pittosporum Family)

Growing to 10 feet high, this useful foundation, filler, or foliage accent plant may also be used as a hedge, container plant, and in arrangements. It may be grown in sun or light shade in most well-drained soils. It has moderate drought and salt tolerance and good wind tolerance.

Plumbago auriculata
BLUE PLUMBAGO, CAPE PLUMBAGO, CAPE LEADWORT
Plumbaginaceae (Plumbago Family)

This is a spreading South African evergreen shrub, mounding to 6 feet high. The flowers are produced much of the year and are used to make leis. It performs best in full sun in a rich, well-drained soil. It has moderate salt and wind tolerance and good heat and drought tolerance. It is useful as a filler or foundation plant, a low hedge, bank cover, and color mass. A darker blue variety, as well as the White Cape Plumbago (*P. auriculata* 'Alba'), are available.

Polyscias fruticosa
PARSLEY PANAX, MING ARALIA
Araliaceae (Panax Family)

From India to southern Polynesia, this evergreen shrub grows to 10 feet high in sun or shade and in most well-drained soils. It is used as an excellent vertical accent, a hedge or screen, or as a container plant for the lanai or interior. It has moderate drought and salt tolerance. The cut stems are used in arrangements and the foliage in leis.

Polyscias guilfoylei 'Crispa'
CURLY LEAF PANAX
Araliaceae (Panax Family)

A slow growing southern Polynesian evergreen to 8 feet high, this slender shrub may be used as a vertical accent or hedge and makes an excellent container plant for the lanai or interior. Cut stems are used in arrangements. It may be grown in sun or shade in most well-drained, moist soils. It has little salt tolerance.

Pseuderanthemum carruthersii

FALSE ERANTHEMUM

Acanthaceae (Acanth Family)

This evergreen shrub grows to 10 feet high and flowers are seen throughout the year. It is native to the southwestern Pacific. While this species is not known to be in cultivation in Hawaiian gardens, there are a number of varieties available to the gardener. All grow well in sun or partial shade, respond to moist, well-drained soils, and have moderate drought and salt tolerance. They make strong color accents and may be used for hedges, screens, and container specimens.

Purple False Eranthemum (*P. carruthersii* var. *atropurpureum*) grows to 8 feet high. Foliage is variegated with irregular patches of dark purple to pink and rose purple flowers.

Yellow-Veined False Eranthemum (*P. carruthersii* var. *reticulatum*) grows to 6 feet high with medium green foliage thickly netted with yellow veins. Flowers are white dotted with purple.

Variegated False Eranthemum (*P. carruthersii* var. *variegatum*) grows to 8 feet high and has dark green foliage marked with irregular patches of grey green and yellow. Flowers are white dotted with purple.

Pseudomussaenda flava

YELLOW MUSSAENDA

Rubiaceae (Coffee Family)

An evergreen shrub from East Africa that grows to 6 feet high, this species is best in moist locations with some wind protection and some shade, but it will tolerate full sun. It is not salt tolerant. Flowering much of the year, it may be used as a specimen plant, in mass plantings, or as a hedge.

Punica granatum 'Nana'

DWARF POMEGRANATE

Punicaceae (Pomegranate Family)

A showy dwarf form of the common pomegranate, this cultivar forms a dense, everblooming shrub to 6 feet high. It thrives in full sun in any well-drained soil and is heat and wind tolerant and partially drought tolerant. It makes an excellent accent, colorful hedge or screen, and does well in a container.

Medium Shrubs

Rhaphiolepis indica
INDIAN HAWTHORN
Rosaceae (Rose Family)

An evergreen shrub from southern China that grows to 5 feet high, this useful plant functions as a ground cover, bank cover, accent, informal hedge, or makes an excellent container plant. It grows well in full sun or light shade in most soils and has moderate drought and salt tolerance as well as good wind tolerance. Stems, flowers, and fruit are used in arrangements.

Rhaphiolepis umbellata var. *integerrima*
YEDDO HAWTHORN, KOKUTAN
Rosaceae (Rose Family)

An evergreen Japanese shrub, this plant attains a 10-foot height and produces white flowers in winter, followed by black fruit. Plant in a sunny place in fertile, well-drained soil. It has good salt, wind, and drought tolerance. This is a basic garden plant for foundation, mass, border, or group plantings or as a potted specimen. Stems, flowers, and fruit are used in arrangements.

Rhododendron indicum
AZALEA, INDICA AZALEA (P)
Ericaceae (Rhododendron Family)

An evergreen from Japan, this plant reaches 6 feet in height with an 8-foot spread. Flower color varies from white to pink, salmon, rose to reds, and lavender. Azalea blooms more heavily during the cool months and sparingly throughout the year. It needs filtered light and an acid, well-drained soil with a mulch over its roots for best results. It may be used as a color mass, foundation, or background planting. It is a popular potted plant and used in arrangements.

Rondeletia odorata
RONDELETIA
Rubiaceae (Coffee Family)

Native to Cuba and Panama, this species grows to 6 feet high and bears bright blooms most of the year. In spite of its name, it is not fragrant. It will thrive in full sun in almost any well-drained soil with moderate moisture. It is a moderate grower and is wind tolerant and partially drought tolerant. It is useful as a color accent, hedge, or screen, or may be used as a potted specimen for a sunny lanai.

Russelia equisetiformis
FIRECRACKER PLANT, CORAL PLANT, LŌKĀLIA
Scrophulariaceae (Snapdragon Family)

This sprawling, weeping Mexican plant grows to 5 feet high and produces bright red flowers all year. It flowers best in full sun in a well-drained soil and has moderate drought tolerance, as well as good wind and salt tolerance. It is good as a color mass in the garden and can also be used in rockeries, window boxes, hanging baskets, or raised containers.

Sanchezia speciosa

SANCHEZIA

Acanthaceae (Acanth Family)

Northern South America is the origin of this evergreen shrub that grows to 8 feet high. It thrives in sun or shade in most well-drained soils and has moderate salt, wind, and drought tolerance. It makes a useful screen or hedge plant, specimen or background plant, and an excellent tubbed color accent for the lanai or interior.

Scaevola taccada

BEACH NAUPAKA, NAUPAKA, NAUPAKA KAHAKAI

Goodeniaceae (Naupaka Family)

Native to beach areas in Hawai'i and throughout the tropical Pacific, this shrub grows to 10 feet high. It thrives in pure sand, tolerates salt, wind, heat, and drought, and is a sand binder. It makes an excellent screen or hedge. The branches and large leaf rosettes are used in arrangements, and the white flowers, fruit, and young leaf rosettes are used in leis.

Sida fallax

'ILIMA

Malvaceae (Hibiscus Family)

There are several forms of this native Hawaiian shrub, which grows to 10 feet high. Flowers appear all year and range in color from yellow to orange to dull red; all are prized by the lei maker. *'Ilima* grows best in full sun in hot, dry places, and provides a color accent in the Xeriscape garden. A low-growing form, *'Ilima Papa*, is listed under Ground Covers (Chapter 1).

Strelitzia reginae

BIRD OF PARADISE, CRANE FLOWER (S)

Strelitziaceae (Bird of Paradise Family)

This slow growing South African evergreen produces flowers all through the year. They are held above the foliage, which may reach 6 feet in height. The foliage and flowers are excellent in arrangements. It blooms best in sun with a rich, moist soil and has moderate wind, salt, and drought tolerance. It is a bright accent specimen and may be used as a hedge, low screen, in groupings, or as a container plant. It is also good near the pool.

Tecoma capensis

CAPE HONEYSUCKLE, I'IWI HAOLE

Bignoniaceae (Catalpa Family)

This rambling evergreen shrub or vine, mounding to 10 feet high, is native to South Africa and produces bright flowers most of the year. It requires good drainage and full sun for best flowering. It is tolerant of most soils, heat, wind, salt air, and some drought. It may be used as a hedge, bank cover, or espalier. Available cultivars include Yellow Cape Honeysuckle (*T. capensis* 'Aurea'), as well as ones with gold- and apricot-colored flowers.

Medium Shrubs

Thunbergia erecta
BUSH THUNBERGIA, KING'S MANTLE
Acanthaceae (Acanth Family)

An evergreen African native to 6 feet high, this shrub carries blue purple flowers most of the year. It makes a good screen, hedge, border, and container plant. Bush Thunbergia has moderate drought, salt, and wind tolerance and grows well in either sun or light shade in most soils with moisture. The cultivar White Bush Thunbergia (*T. erecta* 'Alba') produces white flowers.

Xanthosoma robustum
'APE, COMMON 'APE (P) (S)
Araceae (Aroid Family)

Originating in tropical America, this accent specimen will rapidly reach 10 feet in height, growing in rich, moist soil, in sun or shade. A root barrier should be employed to prevent invasiveness. It requires protection from wind. The flowers, pinkish and sweet smelling, appear during spring and summer. It provides the ultimate tropical accent.

Chapter 4
Large Shrubs

This grouping of plants includes both woody and herbaceous plants that grow 10 feet high or more.

Acalypha wilkesiana
Copper Leaf, Jacob's Coat, Beefsteak Plant (P)
Euphorbiaceae (Euphorbia Family)

A sprawling, dense evergreen plant from Southeast Asia, this colorful accent grows rapidly to 15 feet high. It finds use as a hedge or screen and as a tubbed specimen. It shows its color best in full sun but will tolerate light shade. It has moderate salt and drought tolerance and will grow in almost any moist soil. An interesting cultivar, Picotee Acalypha (*A. wilkesiana* 'Picotee'), displays a unique fringed leaf margin and has similar growth requirements and uses in the landscape.

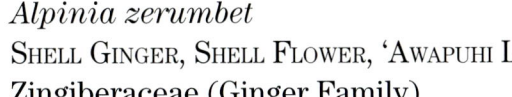

'Picotee'

Alocasia macrorrhizos
Giant Taro, Elephant's Ear, 'Ape (P) (S)
Araceae (Aroid Family)

Growing rapidly to 12 feet high, this is a major garden foliage accent used either singly or massed as a background for lesser tropicals. It may also be used as a tubbed specimen in full sun, shade, or the interior. It responds best in a rich, organic, moist, well-drained soil and needs protection from the wind. It is not salt tolerant.

Alpinia zerumbet
Shell Ginger, Shell Flower, 'Awapuhi Luheluhe
Zingiberaceae (Ginger Family)

This herbaceous plant from Melanesia grows to 12 feet high and produces hanging, fragrant flowers much of the year. It does well with protection from strong winds, in full sun or light shade, in a rich, moist soil. It has moderate salt tolerance. The blooms are excellent for arrangements.

Bixa orellana
Lipstick Plant, Achiote, 'Alaea
Bixaceae (Annatto Family)

This shrub from tropical America is moderately fast growing, up to 30 feet high, bearing clusters of pods variously colored from pinkish red to red or yellow. The pods are prized for use during the holidays for arrangements and wreaths. This evergreen plant can be used as a color specimen, background, enclosure, or hedge. It does well in most soils with good drainage and moisture but needs full sun for best flowering. It has low salt and wind tolerance and moderate tolerance to drought.

Breynia disticha 'Roseopicta'
Snowbush, Sweetpea Bush, Calico Plant
Euphorbiaceae (Euphorbia Family)

Growing to 10 feet high or more, this evergreen shrub from Melanesia performs best in sun in a light, sandy soil. However, it tolerates most soils, light shade, and some drought. It is not salt tolerant. It makes an interesting potted specimen, a colorful screen, or a hedge. It produces suckers from the roots and may become invasive.

Calliandra haematocephala
Red Powderpuff, Lehua Haole
Fabaceae (Bean Family)

A fast growing shrub from Bolivia, this evergreen reaches 16 feet in height with a 20-foot spread. It produces large, showy red or reddish pink inflorescences in fall and winter. It prefers full sun in a rich, well-drained soil, has moderate salt and drought tolerance, and good tolerance to wind and heat. It makes a good background, enclosure, or informal screen planting and may be used as a specimen or accent plant where room permits.

Calotropis gigantea
Crown Flower, Giant Milkweed, Pua Kalaunu (P) (S)
Aesclepiadaceae (Milkweed Family)

An evergreen shrub from India and Southeast Asia, this species grows to 15 feet high, bearing its flowers most of the year. The flowers, both white and lavender, are used in leis. Grown best in full sun, it is an accent specimen or can be used as a hedge or screen. It grows readily in any well-drained soil—even pure sand—and is drought, heat, wind, and salt tolerant.

Carissa macrocarpa
Natal Plum (T)
Apocynaceae (Dogbane Family)

A fast growing South African shrub, reaching 20 feet in height, this tough plant flowers and fruits best in full sun. The red fruits are used to make jelly. Its fragrant flowers appear most of the year. It is drought, heat, wind, and salt tolerant and does well in a variety of soils, including pure sand. It makes an excellent barrier hedge, screen, or windbreak and can be used to make an interesting potted specimen. It is also used in arrangements.

Clerodendrum buchanani var. *fallax*
PAGODA FLOWER, LAUʻAWA
Verbenaceae (Verbena Family)

Originally from Java, this evergreen shrub grows to 12 feet high. Its brilliant flowers are produced much of the year. This color accent shrub may be grown in sun or light shade in most moist soils. It is suitable for mass plantings, as a hedge or colorful screen, and as a container plant.

Clerodendrum inerme
SORCERER'S BUSH
Verbenaceae (Verbena Family)

Reaching 10 feet or more in height, this spreading, useful plant has its origins in the wide area from the shores of the Indian Ocean to Malaysia and to many tropical Pacific islands. It is very salt, heat, and wind tolerant and thrives in any soil, including pure beach sand. Its habitat is just beyond the reach of the waves, but it is equally at home away from the coast. Use it as a bank cover, sand binder, or large screen. It fares well when clipped and can become a formal hedge in situations where other species do not survive.

Clerodendrum myricoides 'Ugandense'
BLUE FLOWERED CLERODENDRUM
Verbenaceae (Verbena Family)

As the name implies, tropical East Africa is home to this 10-foot shrub. Flowering much of the year, its blooms provide a rare color in tropical gardens. It does well in sun or light shade in almost any well-drained soil. Use it as a hedge or screen or as a colorful backdrop for shorter flowering specimens.

Clerodendrum quadriloculare
QUEZONIA, BAGAUAK
Verbenaceae (Verbena Family)

Reaching 15 feet in height, this species from the Philippines can be used as a color accent plant, a hedge or screen, or a tubbed specimen for the lanai. It can also be readily pruned into a small tree for an entry court accent. It thrives in full sun in most moist, well-drained soils with moderate protection from harsh wind. Flowering in fall and spring, its color value is enhanced by the deep purple of its leaf underside.

Codiaeum variegatum
CROTON
Euphorbiaceae (Euphorbia Family)

This evergreen Melanesian shrub reaches 20 feet in height and is prized for its highly variable foliage color, size, and shape. It makes a good container plant for both the lanai and the interior, serves as a hedge or screen, and can be pruned into a small tree. It will take sun or moderate shade in most soils. It has moderate drought and salt tolerance and good tolerance to wind.

Large Shrubs

'Peter Buck'

'Bob Alonzo' 'Floozie'

'Hawaiian Flag' 'Lau Kea'

'Haole Girl' 'Iwao Shimizu'

Cordyline fruticosa
GREEN TI, TI, KĪ
Agavaceae (Agave Family)

Green Ti, a slender plant originally from East Asia, reaches 15 feet in height and is best grown in a rich soil with moisture in light to moderate shade. It has fair salt and wind tolerance. An excellent container plant for the deck or interior, Green Ti may be used as a hedge, a screen, an accent, or a foundation or background plant. The large leaves are used by lei makers and in arrangements. There are hundreds of cultivars, displaying a wide range of plant size and leaf shape, size, and color. Typical are *C. fruticosa* 'Peter Buck,' which grows tall and has large orange red leaves; *C. fruticosa* 'Bob Alonzo,' displaying 12-inch leaves in the orange yellow range; *C. fruticosa* 'Floozie,' with neon red leaves 18 inches long; *C. fruticosa* 'Hawaiian Flag,' with narrow foliage striped with red and yellow; *C. fruticosa* 'Lau Kea,' a slender-leaved cultivar with pale yellow foliage 10 inches long; *C. fruticosa* 'Haole Girl,' with the same color pattern displayed on a shorter, broad leaf; and *C. fruticosa* 'Iwao Shimizu,' a dwarf with small green leaves edged with orange up to only 6 inches long.

Dodonaea viscosa
'A'ALI'I
Sapindaceae (Soapberry Family)

A native Hawaiian evergreen species growing to 24 feet high, this versatile plant finds many uses in the landscape: as a tree, hedge, windbreak, or screen. Its papery fruits, varying in color from white to red to mahogany, are used by lei makers. It is highly wind, drought, and salt air tolerant and thrives in any well-drained soil. It is a moderate grower.

Dracaena fragrans 'Warneckii'
STRIPED DRACAENA
Agavaceae (Agave Family)

This attractive plant from tropical Africa produces many upright stems and grows to 15 feet high. It makes an excellent specimen, thriving in most well-drained soils. It has little salt tolerance and moderate drought tolerance and does best sheltered from wind. It requires light to moderate shade and is an excellent container plant for the lanai or for interior use. It can also be used as a hedge or screen.

Duranta erecta
GOLDEN DEWDROP, DURANTA, PIGEON BERRY (P)
Verbenaceae (Verbena Family)

A useful plant for developing hedges and screens, as well as background and foundation plantings, this tropical American evergreen species grows to 18 feet high with an almost equal spread. Flowering best in full sun in most soils, it displays moderate drought and salt tolerance and good wind tolerance. There is a green and white variegated foliage form available, as well as the White Golden Dewdrop (*D. erecta* 'Alba'). Foliage and fruit are used in arrangements.

Etlingera elatior
Torch Ginger. 'Awapuhi Ko'oko'o
Zingiberaceae (Ginger Family)

Malaysian and Indonesian jungles are home to this herbaceous plant that reaches 20 feet in height. It produces its striking inflorescences on stalks arising directly from the ground; they are used in arrangements. It grows best in a rich, organic soil with moisture and protection from the wind. It may be planted in sun or partial shade and used as a specimen plant or in mass plantings.

Eugenia uniflora
Surinam Cherry, Pitanga
Myrtaceae (Eucalyptus Family)

This tropical American evergreen shrub or small tree attains a height of 25 feet, bearing fragrant flowers followed by edible fruit. It makes an excellent tubbed specimen but can also be used as a hedge or a good informal screen or background plant. Plant in most well-drained soils but in full sun for best flowers and fruit. It is not salt tolerant.

Euphorbia leucocephala
Pascuita, Flor de Niño
Euphorbiaceae (Euphorbia Family)

A Central American shrub that grows to 20 feet high, this species produces masses of small flowers with showy white bracts during late fall and is valued as holiday decor. It makes a good specimen or screening and border plant. It is best planted in full sun in a well-drained soil.

Euphorbia pulcherrima
Poinsettia, Christmas Flower
Euphorbiaceae (Euphorbia Family)

No doubt the most symbolic herald of the Christmas season worldwide, this Mexican and Central American shrub is loved in every Christian household as it flowers during late fall and early winter. There are many cultivars with varying bract colors, ranging from the traditional deep red to pink to white and also double, splotched, and dotted variations. Grow poinsettia in good, well-drained, moist soil. In addition to its popular potted condition, it will become, with judicial pruning, a beautiful small specimen tree to 15 feet high for the entryway or a brilliant winter hedge or screen. Colored bracts may remain for several months.

Ficus microcarpa var. *crassifolia*
Wax Fig, Taiwan Fig
Moraceae (Mulberry Family)

This sprawling evergreen shrub from Asia reaches 12 feet in height. It can be grown in any soil, in sun or shade, and has good salt, drought, and wind tolerance. It is an excellent bank cover or filler and is easily pruned to form a hedge or an interesting containerized specimen for the lanai.

Gardenia taitensis
Tahitian Gardenia, Tiare, Tiare Tahiti
Rubiaceae (Coffee Family)

This Polynesian evergreen shrub or small tree, moderately fast growing to 20 feet high, bears its fragrant flowers from spring to fall. Individually picked flowers last several days and can be used in leis. Tiare may be used for a screen, bank cover, background or foundation planting, or trained as a small tree. It flowers best in sun in any good garden soil. It has moderate wind tolerance and good salt tolerance.

Grewia occidentalis
Lavender Starflower
Tiliaceae (Linden Family)

A fast growing, sprawling evergreen shrub to 10 feet high, starflower originates in South Africa and flowers much of the year, more abundantly in cooler areas. It will do well in hot, dry areas with adequate moisture but is best in sun or partial shade. A useful specimen or container plant, it is also good in mass plantings or espaliered.

Heliconia caribaea
Gold Heliconia
Heliconiaceae (Heliconia Family)

A large plant that grows to 18 feet high, this West Indian native presents its massive blooms during the warm months of the year. Protection from wind and lots of water are a must for successful production. This heliconia grows rapidly in a rich, well-drained loam. There are several cultivars available, including those with deep red, medium red, yellow and green, yellow and red, and gold and yellow bracts. All are excellent and long-lasting in arrangements.

'Lemon Chiffon' 'Madam Pele'

Hibiscus rosa-sinensis
Red Hibiscus, Common Hibiscus, Chinese Hibiscus
Malvaceae (Hibiscus Family)

This evergreen Asian plant that reaches 15 feet in height is valued as an accent, hedge, screen, or background plant. Growing well in sun or light shade in most well-drained soils, it tolerates heat and alkaline soils and has moderate salt and drought tolerance. The blooms are bright red and are exhibited all year, but each bloom lasts only one day. Flowers are used in arrangements. Innumerable cultivars are available with single or double flowers in a range of colors and bicolors from white through pink, coral, yellow, orange, coffee, and lavender to red, with both erect and pendant blooms and with variegated foliage. Stature varies from low shrubs to treelike specimens. Shown are representative cultivars: 'Lemon Chiffon,' 'Madam Pele,' 'Nii Yellow,' 'Higa Yellow,' and 'Snowflake,' with variegated foliage.

'Nii Yellow' 'Higa Yellow' 'Snowflake'

Hibiscus schizopetalus
Coral Hibiscus, Fringed Hibiscus, Aloalo Koʻakoʻa
Malvaceae (Hibiscus Family)

This evergreen shrub of weeping habit from tropical East Africa grows to 15 feet high, bearing its unusual flowers all year. They are used in arrangements. It may be used as a specimen, in mass plantings, or as a screen, hedge, or background planting. It is best in sun or light shade in most well-drained soils but is not salt tolerant. A number of hybrids are available showing pendant, fringed flowers, ranging from pure white, pink, and yellow to red. These are more commonly seen than the parent species. Shown are: 'Pink Butterfly,' 'Butterfly White,' and 'Itsy Bitsy Pink.'

'Pink Butterfly'

'Butterfly White'   'Itsy Bitsy Pink'

Holmskioldia sanguinea
Cup and Saucer, Chinese Hat Plant, Parasol Flower
Verbenaceae (Verbena Family)

A sprawling evergreen shrub from the Himalayan foothills, this species grows to 30 feet high, bearing showy flowers much of the year. Color varies from brick red in the species through gold and orange in hybrids. The Yellow Cup and Saucer cultivar (*H. sanguinea* 'Citrina') is clear yellow. They are best in sun in a well-drained soil, have fair drought and salt tolerance, and good tolerance to wind and heat. A useful color specimen and background or screening plant. Flowers are used in leis.

Juniperus chinensis 'Torulosa'
Hollywood Juniper
Cupressaceae (Cypress Family)

This temperate Asian shrub grows well in Hawaiian gardens, especially in cool upland areas. Reaching 15 feet in height, it forms an unusual accent plant and is seen in gardens of Oriental inspiration or as a potted specimen, screen, hedge, or windbreak. It prefers full sun and will thrive in any well-drained soil. It has good wind tolerance and some salt tolerance. Foliage is used in arrangements.

Leea guineensis
Leea, West Indian Holly
Leeaceae (Leea Family)

This tropical West African evergreen shrub grows to 15 feet high and produces red flower clusters in late spring and summer. It may be grown in sun to dense shade and does well in most well-drained soils. It has low salt and drought tolerance. It is used as a vertical accent, a screen or hedge plant, and is excellent in a container indoors or out. A form with purple red foliage is available.

Ligustrum japonicum
JAPANESE PRIVET (A)
Oleaceae (Olive Family)

A Japanese evergreen shrub that grows to 15 feet high, this species produces white flowers in the spring. It will take sun or partial shade, a wide range of soil conditions, and has good tolerance of coral sand and salt spray. It can be used as a specimen pruned to tree shape, a hedge, screen, or background plant. A cultivar, Round Leaf Privet (*L. japonicum* 'Rotundifolium'), has more compact growth and rounded leaves.

Lysiloma thornberi
FEATHER BUSH
Fabaceae (Bean Family)

This evergreen plant from dry parts of the southern United States takes heat and drought and has good wind and moderate salt tolerance. At 15 feet high, it makes a good informal background shrub or it may be trained as a small tree. It is at its best in sun in a well-drained soil.

Malvaviscus penduliflorus
TURK'S CAP, FIRECRACKER HIBISCUS, PAHŪPAHŪ
Malvaceae (Hibiscus Family)

This tropical American evergreen shrub, up to 15 feet high, flowers all year. The red, pink, or white flowers are used in lei making. Turk's Cap is at its best in sun in a moist, well-drained soil. It may be used as a specimen plant, an informal hedge or screen, and is readily pruned into a "standard."

Murraya paniculata
MOCK ORANGE, CHINESE BOX, ALAHEʻE HAOLE (A)
Rutaceae (Citrus Family)

This moderately slow growing shrub or small tree reaches 25 feet in height and is native to an area from India to the Philippines. Its fragrant flowers appear several times during the year. Foliage and flowers are used in arrangements and in leis. It is widely used as a hedge, screen, or windbreak. It is best in sun, but it tolerates some shade and thrives in a well-drained soil. It has only fair salt and drought tolerance but good wind tolerance.

Mussaenda x 'Doña Luz'
DOÑA LUZ MUSSAENDA
Rubiaceae (Coffee Family)

A 12-foot tall Philippine cultivar, this spectacular shrub flowers heavily from spring through fall, making a strong color accent. It can also be used as a hedge, in mass plantings, and as a containerized specimen. It prefers sun in a moist location in good soil and appreciates protection from wind. It is not salt tolerant.

Mussaenda erythrophylla 'Doña Trining'
DOÑA TRINING MUSSAENDA
Rubiaceae (Coffee Family)

A cultivar from tropical West Africa that grows to 15 feet high, this plant requires the same cultural care and offers similar landscape applications as its relative (above) from the Philippines.

Mussaenda philippica 'Doña Aurora'
DOÑA AURORA MUSSAENDA
Rubiaceae (Coffee Family)

A cultivar from the Philippines, this shrub grows to 15 feet high. It has the same cultural requirements and flowering season as 'Doña Luz.'

Mussaenda x 'Queen Sirikit'
QUEEN SIRIKIT MUSSAENDA
Rubiaceae (Coffee Family)

This is another cultivar from the Philippines that grows to 10 feet or more in height and has the same cultural requirements and landscape uses as 'Doña Luz.'

Nerium oleander
OLEANDER, COMMON OLEANDER, 'OLIANA (P)+ (S)+
Apocynaceae (Dogbane Family)

A dense, fast growing shrub from the region extending from the Middle East to Japan, this evergreen grows to 30 feet high. There are many cultivars available with more compact growth habit, variegated foliage, and a range of fragrant, showy flowers from red, pink, salmon, and pale yellow to white, in both single and double forms. This highly drought and heat tolerant plant should be used in sun for best flowering. It does well in almost any soil and has good wind and salt tolerance. Excellent for an informal sound, wind, or visual barrier, it is also useful as a tubbed specimen or trained as a small tree.

Philodendron bipinnatifidum
TREE PHILODENDRON (P) (S)+
Araceae (Aroid Family)

This South American species develops a stout trunk up to 12 feet high supporting a dense head of foliage. The Tree Philodendron is valued in the landscape as an accent, major ground or bank cover, or as a container plant for the lanai or interior. It grows well in sun or shade and likes moisture and a rich soil. The foliage looks best when protected from harsh winds. It has minor salt tolerance but no drought tolerance.

Pittosporum tobira
JAPANESE PITTOSPORUM, TOBIRA
Pittosporaceae (Pittosporum Family)

An evergreen from China and Japan that grows to 25 feet high, this tough plant thrives in sun or shade in most soils with good drainage. It has good wind, salt, and drought tolerance. Plant this pittosporum as an informal screen, windbreak, border, or hedge. It is also effective in containers and highly useful when trained as a small tree. Branches may be used in arrangements.

Polyscias filicifolia 'Golden Prince'
GOLDEN PRINCE ARALIA
Araliaceae (Panax Family)

This evergreen shrub from Southeast Asia and South Pacific islands grows to 15 feet high. It finds good landscape use as a specimen, vertical accent, hedge, screen, container plant, or to frame a doorway or view. It develops its best color in strong light in a rich, moist, well-drained soil. It is used in arrangements.

Polyscias guilfoylei
PANAX, COMMON PANAX
Araliaceae (Panax Family)

This fast growing, evergreen, fastigiate shrub from southern Polynesia attains a height of 20 feet and is the perfect hedge in narrow confines. Planted in sun or light shade in almost any well-drained soil, it is also useful as a screen or background subject. It has moderate salt, wind, and drought tolerance. There are plants available with solid green foliage.

Polyscias scutellaria 'Balfouriana'
WHITE-EDGE BALFOUR ARALIA
Araliaceae (Panax Family)

From Java, this 15-foot-tall, slender evergreen shrub finds excellent use as a container plant, accent, hedge, or screen and in arrangements. It performs best in full sun or light shade, in a well-drained, rich soil with regular watering and feeding. It is wind tolerant and withstands salt air in a sheltered position.

Schefflera arboricola
DWARF BRASSAIA
Araliacea (Panax Family)

An irregular, shrubby evergreen vine from Taiwan that grows to 20 feet or more, this species may be used in sun or shade in most soils. It has good wind and drought tolerance and makes a good hedge, enclosure, or screen plant and an excellent container plant for the deck, lanai, or interior. It can be trained as a dense espalier or readily pruned to maintain a shrubby habit.

Tabernaemontana divaricata
CREPE JASMINE, PAPER GARDENIA
Apocynaceae (Dogbane Family)

Fast growing, this evergreen shrub from northern India grows to 15 feet high, producing its flowers throughout the year. It is used in the shrub border, as a background or foundation planting, or trained as a small tree. It grows well in sun or partial shade in many different soils. It has moderate drought and wind tolerance but is not tolerant of salts. The Butterfly Gardenia (*T. divaricata* 'Flore Pleno') makes a more attractive shrub with its clusters of double flowers, pure white and lacy, and its larger leaves, wavy and linear.

Chapter 5
Small Trees

This group includes those plants that grow into trees with either single or multiple trunks and reach heights of 15 to 30 feet. These are especially useful in Hawai'i, with our smaller residential lots and limited space for street plantings.

Bauhinia monandra
PINK BAUHINIA, ST. THOMAS TREE
Fabaceae (Bean Family)

A fast growing, deciduous, open-crowned species from Myanmar, this tree grows to 30 feet high and bears showy flowers in the spring and summer. A good general garden or specimen tree for most well-drained soils, this tree has good drought tolerance but little salt tolerance.

Bauhinia tomentosa
YELLOW BAUHINIA
Fabaceae (Bean Family)

This is an evergreen tree from Africa to China with a moderate growth rate, attaining a height of 20 feet. The bright flowers are produced much of the year. It is good for residential or street plantings in hot, dry locations and does best in sun in most soil conditions. It has moderate wind tolerance but is not salt tolerant.

Brugmansia x candida
ANGEL'S TRUMPET, NĀNĀHONUA (P)
Solanaceae (Potato Family)

A fast growing South American tree up to 20 feet high, this evergreen plant has large, fuzzy, dull green leaves and large, fragrant blooms through summer and fall. In full sun or partial shade in moist soils, it makes a spectacular specimen or hedge. It requires little care once established but has little tolerance to wind and salts. Forms with pale apricot and yellow flowers are available. It is better in cooler areas.

Bucida molineti
DWARF GEOMETRY TREE, SPINY BLACK OLIVE (T)
Combretaceae (Combretum Family)

A slow growing tree from the Bahamas with a strong horizontal branching habit, this attractive plant may reach 25 feet in height. This adaptable plant will grow almost anywhere except under extremely salty conditions or dense shade. Its small greenish flowers are insignificant. It makes an excellent specimen or container plant, a good rock garden plant, a bonsai subject, or in the interiorscape.

Caesalpinia pulcherrima
DWARF POINCIANA, PRIDE OF BARBADOS, 'OHAI ALI'I (P) (T)
Fabaceae (Bean Family)

This fast growing evergreen tree, 15 feet high, carries its bright flowers much of the year. From the West Indies, it is best in full sun in most soils and has good heat, drought, and salt tolerance. A yellow form, Yellow Dwarf Poinciana (*C. pulcherrima* forma *flava*), is available, as well as a salmon-colored flower form. Use this adaptable species as an accent, hedge, screen, or specimen. All color forms are used in making leis.

Callistemon citrinus
RED BOTTLEBRUSH, CRIMSON BOTTLEBRUSH
Myrtaceae (Eucalyptus Family)

An Australian evergreen up to 30 feet high, this colorful species produces its brushlike flowers on drooping branches periodically throughout the year. It grows best in full sun, will perform in most well-drained soils, is drought and wind tolerant, and has moderate salt tolerance. Use this Aussie as a specimen, color accent, or screening plant. There is a range of available cultivars.

Callistemon rigidus
STIFF BOTTLEBRUSH
Myrtaceae (Eucalyptus Family)

A slow growing, dense, stiffly upright tree reaching 20 feet in height, this evergreen Australian species has aromatic foliage and showy, brushlike flowers that appear periodically. Use it as a specimen plant, a screen, or a hedge or for windbreak plantings in dry areas. Best planted in full sun, it does well in most soils and has good drought, salt, heat, and wind tolerance.

Clusia rosea
AUTOGRAPH TREE, SCOTCH ATTORNEY, COPEY
Clusiaceae (Mangosteen Family)

This round-headed evergreen tree up to 30 feet high is native to the Florida Keys and West Indies. The showy pink or white flowers are produced in winter and spring and are followed by applelike green pods that are used by arrangers either fresh or dried. It has excellent drought, salt, and wind tolerance. Use it as a specimen, a container plant, for espaliering, screening, or as a windbreak. A variegated cultivar is known.

Coccoloba uvifera
Sea Grape
Polygonaceae (Buckwheat Family)

This tropical American evergreen tree grows to 30 feet high and has ornamental bark and foliage. The clusters of small, fragrant, white female flowers are followed by edible fruit used to make jelly. It has excellent salt and wind tolerance and may be used as a specimen or for enclosure, screening, or windbreak plantings. Grow in full sun or light shade. It is adapted to most soil conditions, including pure beach sand. Male trees may be grown from cuttings to avoid fruit production.

Coffea arabica
Coffee, Arabian Coffee
Rubiaceae (Coffee Family)

The coffee plant of commerce is also a useful small tree or large shrub for the landscape. Of tropical African origin, this evergreen grows to 20 feet high, bearing white, fragrant flowers and useful red coffee beans. It does best in a moist, well-drained, fertile soil with protection from harsh winds. It makes a good interior plant, a hedge, or a screen.

Cordia sebestena
Kou Haole, Geiger Tree
Boraginaceae (Borage Family)

A dense, round-headed tree up to 25 feet high, this slow growing species from the Caribbean islands and tropical America displays clusters of showy flowers much of the year. It is best planted in sun but will tolerate partial shade and will grow in most well-drained soils. It has good wind, salt, and drought tolerance. Use it as a garden specimen or as an avenue or street tree.

Dracaena marginata
Madagascar Dragon Tree, Money Tree
Agavaceae (Agave Family)

This 25-foot Madagascan tree is usually seen as a potted specimen for the lanai or the interiorscape. It forms a striking, round-canopied accent in the general garden in full sun when permitted to grow untrimmed. Flowers are white and appear in winter. It will grow in almost any soil and is highly drought and wind tolerant. Several cultivars are available with green, white, and pink leaf variations. There is a large-leaved form.

Erythrina crista-galli
Common Coral Tree, Cock's Spur Coral Tree (P)
Fabaceae (Bean Family)

This rapidly growing, semievergreen South American tree may reach 25 feet in height. Flowers, ranging in color from red to pinkish red, are shown most of the year. It makes a good specimen for the lawn or garden in sun or some shade in most well-drained soils. It requires little water or attention once established. Flowers can be used in lei making.

Small Trees

Euphorbia cotinifolia
Red Spurge, Hierba Mala (P) (S)
Euphorbiaceae (Euphorbia Family)

This evergreen tree from Mexico to northern South America has inconspicuous flowers but showy foliage. It is best grown in sun in a well-drained soil, where it will grow to 20 feet high, making a useful garden accent, screen, or container specimen. It has good drought tolerance but does not tolerate salt or strong wind.

Guaiacum officinale
Lignum Vitae
Zygophyllaceae (Lignum Vitae Family)

This very slow growing, bushy evergreen tree, originating in Central America and the West Indies, produces masses of lavender blue flowers in the late spring. Its fruits are used in arrangements. In sun or partial shade and adapted to a wide range of soils, it is useful in the entry court or as an accent or screen in the landscape. Lignum Vitae does well in a container. It tolerates drought and wind but not salt.

Harpullia pendula
Tulipwood
Sapindaceae (Soapberry Family)

This attractive, round-headed evergreen tree up to 25 feet high bears clusters of inconspicuous flowers followed by attractive fruits. This excellent Australian native may be used in a courtyard, to shade a lanai, or as a street tree. It thrives in sun or partial shade in most soils. Fruit clusters are used in arrangements.

Heritiera littoralis
Looking Glass Tree
Sterculiaceae (Cacao Family)

A spreading evergreen tree up to 30 feet high or more, this unusual species comes from eastern Africa and South Pacific islands. Grey leaves, silvery underneath, are used in arrangements, as are the keel-shaped fruits. It has good salt tolerance, likes full sun, will grow in most soils, and has some resistance to wind. Use it as a screen or tall hedge.

Hibiscus arnottianus subsp. *punaluuensis*
Koki‘o Ke‘oke‘o
Malvaceae (Hibiscus Family)

Native to O‘ahu's Ko‘olau Range, this hibiscus grows to 30 feet high, bearing masses of large, fragrant flowers all year. It responds to rich, moist, well-drained, fertile soils. Use this beautiful plant as a small entryway tree or as an accent in the general landscape. It makes a dense, tough hedge, windbreak, or screen and is a moderately rapid grower. Flowers are used by lei makers.

Hibiscus tiliaceus
Hau, Tree Hibiscus
Malvaceae (Hibiscus Family)

Found worldwide along tropic shores and rivers, *Hau* is a highly variable and useful species. In Hawai'i, where it is possibly native, *Hau* becomes a rambling, almost vinelike specimen requiring heavy pruning to produce an upright shape. Growing to 30 feet, it can be used to cover a strongly supported arbor. It is fast growing and evergreen with good salt, wind, and heat tolerance and moderate drought tolerance. Its single, one-day, bright yellow flowers are used in leis. There are semidouble forms. Others, also erect, bear round or tripointed green or mahogany red foliage; green and white variegated foliage; and green and white and pink variegated foliage. These are useful for screening and as color accents. A variety from French Polynesia, *H. tiliaceus* var. *sterile*, grows erect to 40 feet high and can be used for screening and as a street tree.

H. tiliaceus var. sterile

Jatropha integerrima
Rose-Flowered Jatropha, Peregrina (P)+ (S)+
Euphorbiaceae (Euphorbia Family)

This bushy evergreen tree up to 20 feet high comes from the West Indies and flowers throughout the year. It is used in sun or partial shade in a fertile, well-drained soil. Train it as a small specimen tree, use in an informal border or screen, or place it in a container for color accent on the lanai.

Metrosideros polymorpha
'Ōhi'a Lehua
Myrtaceae (Eucalyptus Family)

A highly variable, slow growing native plant is seen in nature from a shrub to tall tree. The leathery foliage is variable. The flowers are commonly red but are also seen in shades of orange, pink, white, and yellow, produced throughout the year. It is best in full sun in a well-drained soil. *'Ōhi'a Lehua* has good wind and drought tolerance but only moderate salt tolerance. Flowers, *liko*, and foliage are used for arrangements and leis.

Morinda citrifolia
Noni, Indian Mulberry
Rubiaceae (Coffee Family)

One of the heritage species, *Noni* possibly originated in Southeast Asia. It is an extremely tough, small tree up to 15 feet high with wind, salt, and drought tolerance. Use it as a hedge, a windbreak or screen, or prune it as an accent in the garden. A word of caution: *Noni* fruit, highly regarded for its medicinal values, emits a rather regrettable odor when fully ripe. Plant it downwind.

Myoporum sandwicense
Bastard Sandalwood, Naio
Myoporaceae (Naio Family)

This moderately slow growing native Hawaiian plant reaches 30 feet in height and requires full sun and a well-drained soil. It has good wind, salt, and drought tolerance. *Naio* makes an excellent windbreak, hedge, screen, or specimen.

Nolina recurvata
Pony Tail, Beaucarnea
Agavaceae (Agave Family)

Deserts of northern Mexico are the native habitat of this popular woody succulent that attains a height of 30 feet. It is slow growing, demands full sun for best growth, and needs a well-drained soil. It is drought and wind tolerant and moderately salt resistant. It makes a good potted or tubbed specimen and a striking accent in the landscape. Dried leaves are used in arrangements.

Pandanus tectorius
Hala, Screwpine
Pandanaceae (Screwpine Family)

This round-headed evergreen tree up to 30 feet high is found on many Pacific islands. It is an excellent shade tree and windbreak for the beach garden, displaying good wind and salt tolerance, and it can be used in containers for the lanai or indoors. It thrives in any well-drained soil, even pure sand. Female plants develop large pineapple-like fruit used in arrangements. Fruit segments, as well as the papery, fragrant white male flower, are used in lei making.

Pittosporum hosmeri
Hōʻawa
Pittosporaceae (Pittosporum Family)

This native species grows to 30 feet high, with a dense canopy of leaves. Flowers are plentiful, white, and highly fragrant at night. Plant *Hōʻawa* in full sun in good, well-drained soil. Use it as a courtyard tree, to frame an entry, as a screen or accent, or as a mass in the general garden plan.

Psidium cattleianum
Purple Strawberry Guava, Cattley Guava, Waiawī ʻUla ʻUla
Myrtaceae (Eucalyptus Family)

A dense, evergreen Brazilian tree up to 20 feet high, this attractive species can be used as a specimen in confined areas and in containers, as a hedge, or as a screening plant. It has moderate drought and salt tolerance and good wind tolerance. Plant it over a ground cover or shrubs, which will absorb the edible red fruit.

Psidium cattleianum forma *lucidum*
YELLOW STRAWBERRY GUAVA, WAIAWĪ
Myrtaceae (Eucalyptus Family)

This loosely branched Brazilian evergreen species up to 30 feet high shows good wind tolerance and moderate salt and drought tolerance. It is a useful specimen tree, makes a loose screen, hedge, or windbreak planting, and may even be developed as a clipped hedge. Plant it over a ground cover that will absorb the crop of edible yellow fruit. Its bark is decorative.

Psydrax odorata
ALAHEʻE
Rubiaceae (Coffee Family)

This Hawaiian native displays dark, glossy foliage and blooms periodically through the year. It is drought and wind tolerant and quite salt tolerant. In good, well-drained soil with regular watering and feeding, it will grow to 30 feet high, forming a handsome evergreen specimen. Use *Alaheʻe* as an entry or courtyard tree or mass it as a hedge, windbreak, or screen. Lei makers prize the fragrant flowers.

Punica granatum
POMEGRANATE, POMEIKALANA
Punicaceae (Pomegranate Family)

Iran is probably the original home of this 20-foot tree, now grown worldwide for its bright red flowers and edible fruit. Best grown in full sun in most soils, it is tolerant of heat, alkaline soils, and drought but is not tolerant of salt air. It may be used as a small specimen tree, for mass plantings, or as a tub plant for deck or lanai.

Schefflera elegantissima
FALSE ARALIA
Araliaceae (Panax Family)

An evergreen tree from New Caledonia that reaches 25 feet in height, this species is valued for its attractive foliage in the partially shaded landscape. Juvenile foliage is quite linear, while mature leaves are broader. Use it as an accent specimen in the entry court. It is a good container plant for interiors and the shaded deck or lanai. Plant it in a good soil mix and keep it well watered.

Schinus terebinthifolius
CHRISTMAS BERRY TREE, WILELAIKI
Anacardiaceae (Cashew Family)

A Brazilian evergreen up to 30 feet high, this species can be readily shaped as a shade or accent tree or for use as a hedge, screen, or windbreak. Tolerating almost any well-drained soil, this versatile plant grows easily in hot and dry windy areas exposed to strong salt air. Although its red berries are used for holiday decorating, gardeners are wise to select only male plants, which will avoid eliminating its plentiful seedlings.

Senna surattensis
SCRAMBLED EGGS, KOLOMANA, KALAMONA
Fabaceae (Bean Family)

This fast growing evergreen tree up to 25 feet high comes from tropical Asia, Australia, and Polynesia. It tolerates wind and salt, is highly drought tolerant, and thrives in any well-drained soil. It can be used in mass or screening plantings for color, as a street or boulevard tree, and as a shade tree for the small garden.

Stemmadenia litoralis
LECHOSO
Apocynaceae (Dogbane Family)

This handsome evergreen tree up to 20 feet high is native to Central America. It performs best in full sun in most garden soils as a specimen tree for the small garden or in mass plantings in larger scenes. It requires regular watering for optimum results.

Strelitzia nicolai
GIANT BIRD OF PARADISE
Strelitziaceae (Bird of Paradise Family)

This "bird" produces a treelike clump up to 30 feet high. The huge flowers are produced most of the year. This South African is best grown in sun in a moist, rich soil. It tolerates moderate salt, wind, and sandy soils. This is a dramatic specimen or can be planted as a screen and can be used as a tubbed plant for the lanai or well-lighted interiorscape. Its flowers and foliage are used in arrangements.

Tabebuia aurea
SILVER TRUMPET TREE
Bignoniaceae (Catalpa Family)

A slow growing Brazilian evergreen tree up to 25 feet high, this popular tree bears flowers varying from golden yellow to pale yellow and foliage varying from gray green to silver green. It flowers in late spring and summer. It flourishes in full sun in a fertile soil but will thrive in almost any circumstance. It has good drought tolerance and moderate tolerance to wind. This is a highly useful specimen for the garden or as a street tree, and it can also be used as a tubbed specimen for the lanai.

Tabebuia berteroi
HISPANIOLAN ROSY TRUMPET TREE
Bignoniaceae (Catalpa Family)

Native to the island of Hispaniola in the Caribbean, this 30-foot species is popular for use as an accent specimen. It can be employed as an open screen or windbreak. A rather rapid grower, it has found good use in new developments where it will persist to become a substantial landscape accent. Thriving in almost any well-drained soil, it is moderately wind and drought tolerant.

Thevetia peruviana
BE-STILL TREE (P)+
Apocynaceae (Dogbane Family)

A tropical American evergreen tree up to 30 feet high, this open-canopied species produces showy flowers all year. It is best grown in sun or partial shade in most well-drained soils. It is best in hot, dry locations and can be used as a specimen, for a screen planting, windbreak, or informal hedge. Forms with white and salmon pink flowers are available.

Tournefortia argentea
BEACH HELIOTROPE, TAHINU
Boraginaceae (Borage Family)

A fast growing, spreading tree up to 25 feet high, this native of coastal areas of tropical Asia and the Pacific is an excellent beach tree for shade and windbreak. It is moderately drought tolerant and totally wind and salt tolerant. It grows only in full sun in any soil, including pure beach sand.

Yucca elephantipes
SPINELESS YUCCA, GIANT YUCCA
Agavaceae (Agave Family)

This is a slow growing, somewhat succulent Mexican evergreen tree up to 30 feet high, bearing showy clusters of flowers in the spring and fall. It is best planted in full sun in any good, well-drained soil. It displays good drought and wind tolerance and moderate salt tolerance. It is an outstanding accent or specimen plant and takes well to container growing. A form with green and white variegation is available.

Small Trees

Chapter 6
Medium Trees

This group of plants ranges in height from 30 to 50 feet. It defines the upper limits of our gardens, providing shade, shelter, and color.

Acacia confusa
Formosa Koa
Fabaceae (Bean Family)

An evergreen tree up to 50 feet high, this species is native to the Philippines and Taiwan. Its showy flowers appear in spring and summer. It is a useful specimen tree for lawns, roadside plantings, or in groupings. It has good wind tolerance and moderate salt and drought tolerance. A nitrogen fixer, it will thrive even in poor soils. Foliage, bruised in strong winds, may emit a somewhat unpleasant odor. Plant it downwind.

Acacia koa
Koa
Fabaceae (Bean Family)

This endemic tree grows rapidly to 50 feet high, with an open, wide-spreading canopy. It cannot be recommended for landscape use on Oʻahu due to an as yet unknown pathogen fatal to even young, vigorous trees. It is highly suited to other locations, however, where it makes an ideal shade tree or specimen. A nitrogen fixer, it thrives in poor soils.

Adenanthera pavonina
False Wiliwili, Circassian Bean, Red Sandalwood
Fabaceae (Bean Family)

A Southeast Asian semievergreen tree up to 50 feet high, this species finds good use in the landscape as a shade or street tree. It is rapid growing and produces a light, open crown. It has good wind tolerance and is partially drought tolerant. Its small, bright red seeds are used to make jewelry.

Aleurites moluccana
Kukui, Candlenut Tree (P)
Euphorbiaceae (Euphorbia Family)

Fast growing to 35 feet high, this evergreen tree is native to Polynesia and southern Asia. It is used as a specimen or massed in a grove. It works best in full sun in soil with good drainage. Leaves, flowers, and seeds are used in leis.

Azadirachta indica
NEEM TREE, NIM TREE
Meliaceae (Mahogany Family)

Native to India and Sri Lanka, this evergreen tree grows rapidly to 50 feet high, with fragrant, white flowers. It tolerates poor, marginal soils, drought, and salt water and has moderate wind tolerance. Its dense foliage makes it an excellent shade, screen, or windbreak tree.

Bauhinia binata
ALIBANGBANG
Fabaceae (Bean Family)

An evergreen tree up to 35 feet high with a weeping habit, this useful tree is from the Philippines. It is a moderate grower and makes an excellent garden specimen or street tree in full sun or light shade in most well-drained soils. It has good wind tolerance and moderate tolerance to salt and drought.

Bauhinia x *blakeana*
HONG KONG ORCHID TREE
Fabaceae (Bean Family)

Growing to 40 feet high, this semievergreen tree is a hybrid from China. The fragrant flowers are produced throughout the year, but they are more abundant in the cooler months. It does not produce seed pods. It makes a good garden specimen or street tree in sun or light shade in most well-drained soils. It is not salt tolerant but has moderate tolerance to drought and wind.

Bauhinia variegata
ORCHID TREE
Fabaceae (Bean Family)

A dense, round-headed evergreen tree growing rapidly to 40 feet high, this species from India and China produces its bright flowers much of the year. It is used for shade, as a flowering accent for the garden, or as a street tree. It is not salt tolerant but has moderate tolerance to wind and drought. Flower color varies from white to light and dark purple.

Bucida buceras
GEOMETRY TREE, JUCARO, BLACK OLIVE TREE (T)
Combretaceae (Combretum Family)

This bushy, slow growing evergreen tree grows to 50 feet high. Native to Florida, the West Indies, and Central America, it develops an odd forking but attractive branching habit. The Geometry Tree has excellent wind and salt tolerance and moderate drought and shade tolerance. It is a good shade or street tree and makes a dense, high screen.

Cassia fistula
GOLDEN SHOWER, YELLOW SHOWER, INDIAN LABURNUM
Fabaceae (Bean Family)

Growing to 40 feet high, this Indian tree is prized for its long, pendant flower clusters, which are borne from March to September. It makes a good specimen for the garden, in parks, or as a street tree. It has good drought tolerance and flowers best in hot, dry areas. The seed pods can present a litter problem and have an unpleasant odor.

Cassia grandis
PINK SHOWER, CORAL SHOWER
Fabaceae (Bean Family)

A fast growing tree up to 50 feet high, this species from tropical America bears its abundant flowers in early spring. It finds use as a color accent, shade street, or park tree. It requires sunny locations. It is partially drought tolerant but flowers best in a well-watered situation. The seed pods can present a litter problem.

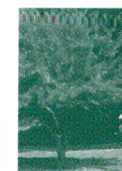

Cassia javanica
PINK AND WHITE SHOWER
Fabaceae (Bean Family)

A round-headed deciduous tree from Indonesia, this species grows to 35 feet high, bearing masses of blossoms in the spring. It makes a good color accent, shade, street, or park tree. It has good wind tolerance and moderate salt tolerance. The seed pods can present a litter problem.

'Lunalilo Yellow'

'Queen's Hospital White'   'Wilhelmina Tenney'

'Nii Gold'

Cassia x *nealiae*
RAINBOW SHOWER
Fabacaeae (Bean Family)

Originating in Hawai'i (a hybrid of *C. fistula* and *C. javanica*), this is an excellent color accent or specimen for the garden or park, and a useful street tree. It is a moderately rapid grower that reaches 40 feet in height, displays good drought tolerance and moderate wind tolerance, but does not tolerate salts. It thrives in almost any well-drained soil. Several named cultivars are available with a range—or rainbow—of flower colors from pale yellow, yellow, and golden yellow to orange and cerise.

'Lunalilo Yellow' has flower buds that open a bright yellow orange from May to September. The petals will fade to a bright yellow with age and are sweetly fragrant. A few seed pods may be produced.

'Queen's Hospital White' has flowers that open pale yellow, fading rapidly to very light yellow to white. Flowering from March through August, it usually produces a few seed pods.

'Wilhelmina Tenney' produces spectacular clusters of flowers from March through September and has not been observed to set seed pods. The petals open with a deep cerise color outside, fading to paler shades with age. The insides of the petals are yellow. This cultivar has been adopted as the official tree of the city of Honolulu.

'Nii Gold' is a sport of 'Wilhelmina Tenney,' displaying new flower buds of a deep gold fading to strong yellow. Its month-long flowering period is highly variable, occurring anytime from spring to fall. It rarely sets seed.

Chorisia speciosa
FLOSS SILK TREE (T)
Bombacaceae (Bombax Family)

A fast growing South American deciduous tree up to 50 feet high, this species produces flowers ranging from white to pink in winter before the onset of spring foliage. It makes a good color accent specimen tree for parks. It has good drought tolerance and moderate salt and wind tolerance.

Chrysophyllum oliviforme
SATINLEAF
Sapotaceae (Sapodilla Family)

Native to tropical America, this dense evergreen tree reaches 35 feet in height. It is prized for its use as a unique foliage-color accent in residential landscapes and parks or as a street tree. It makes a good high screen and is moderately wind and drought tolerant. Its edible fruits are purple. It is not salt tolerant.

Citharexylum spinosum
FIDDLEWOOD
Verbenaceae (Verbena Family)

A popular, fast growing West Indian tree, this species is used as a shade or street tree where quick results are desired. It grows to 50 feet high in either sun or light shade, in both wet and dry conditions. It is generally not recommended for permanent plantings as it has a tendency to produce suckers and water sprouts and is subject to breakage in strong winds.

Cochlospermum vitifolium
BUTTERCUP TREE
Bixaceae (Annatto Family)

This fast growing, open-branched deciduous tree up to 40 feet high is native to tropical America. Its brilliant flowers are produced in winter and early spring before new foliage appears. It makes a suitable street, framing, or specimen tree and has good drought tolerance and moderate wind and salt tolerance. Brazilian Rose (*C. vitifolium* 'Plenum') from Puerto Rico is more commonly seen, displaying its showy, roselike double flowers that measure up to 5 inches across.

'Plenum'

Colvillea racemosa
COLVILLEA
Fabaceae (Bean Family)

An open deciduous tree up to 50 feet high, this Madagascan species produces spectacular 18-inch-long racemes of flowers in fall when most trees are colorless. This moderately slow growing plant has good drought and wind tolerance and moderate tolerance to salt. It is an outstanding flowering color specimen and shade tree in the home landscape and can be used as a street tree.

Conocarpus erectus
SILVER BUTTONWOOD, SILVER BUTTON MANGROVE
Combretaceae (Combretum Family)

This is the popular silver-leaved form of the rarely seen buttonwood. A moderate grower, it may reach 35 feet in height. Its sinuous trunk with rough, furrowed bark is attractive. It has excellent salt and wind tolerance. While it grows best in full sun, it tolerates light shade. It is used in beach gardens for shade, screening, and as a windbreak and also does well inland. It is a good container plant, a brilliant accent for the nightscape, and can be readily pruned into a dense hedge. It is native to tropical America and the West Indies.

Cordia subcordata
Kou
Boraginaceae (Borage Family)

Kou is a dense, round-canopied evergreen tree from the seashores of East Africa to Polynesia. Growing to 35 feet high, it has moderate wind, drought, and salt tolerance. It is best in sun as a shade tree for home gardens and parks. Its flowers, produced much of the year, are used by lei makers.

Crescentia cujete
Calabash Tree, Laʻamia
Bignoniaceae (Catalpa Family)

Native to tropical America, this dense, round-headed evergreen tree grows to 40 feet high. Growing best in hot, dry places, it requires full sun and thrives in most well-drained soils. It has good wind tolerance and moderate drought and salt tolerance. It is useful as a shade tree or specimen where its unique branching habit can be seen. Its rough bark makes a good home for epiphytes. Its large, spherical fruits are used in arrangements, either in the green or dried state.

Cupressus sempervirens
Italian Cypress
Cupressaceae (Cypress Family)

A stately fastigiate tree from the eastern Mediterranean, this tree figures in classical mythology. Because of its slow growth rate, this large tree is considered in the medium range for local gardens. Italian Cypress is drought and wind tolerant and will thrive in areas of mild salt air. It is a popular vertical accent in the landscape and is used to form a slender screen or enclosure. Its foliage can be used in arrangements.

Delonix regia
Royal Poinciana, Poinciana
Fabaceae (Bean Family)

This wide-spreading, rapidly growing deciduous tree from Madagascar attains a height of 40 feet, displaying its brilliant flowers from late spring to fall. Flower color ranges from deep red through red, orange, and clear yellow. It is best in sun and is adapted to a wide range of soils. It has good drought tolerance and moderate tolerance to wind and salt. The Poinciana is a spectacular free-standing color specimen or shade tree for the garden, street, or park. Wide-spreading roots require ample ground space.

Elaeodendron orientale
FALSE OLIVE
Celastraceae (Bittersweet Family)

False Olive, with its dense foliage clothing a rather broad columnar form 40 feet high, makes an excellent screen or windbreak. It is a good general landscape tree for the garden, in parks, and as a street tree. Seedling trees, which display slender juvenile foliage strongly marked with mahogany red, are prized as container specimens. This tree has good wind tolerance and moderate tolerance to drought and salt. It is native to Mauritius.

Erythrina sandwicensis
WILIWILI (P)
Fabaceae (Bean Family)

An extremely drought tolerant native Hawaiian deciduous tree, *Wiliwili* provides a 35-foot high selection for the Xeriscape. It bears masses of flowers that vary from white to chartreuse, coral, orange, and red, as well as bicolors. Flowering precedes the flush of new foliage, although in cultivation, with minimal watering, leaves remain on the tree. Use *Wiliwili* as a color specimen. The blossoms and seeds are valued in the making of leis.

Fagraea berterana
PUA KENIKENI
Loganiaceae (Strychnine Family)

A round-headed to spreading evergreen tree up to 40 feet high, this species from the southern Pacific Islands and Queensland bears highly fragrant flowers throughout the year, opening white and turning yellow. They are used in leis and perfumes. Humid, moist, sunny locations produce best growth. *Pua Kenikeni* makes a good specimen tree or may be used as a screen.

Ficus lyrata
FIDDLE LEAF FIG
Moraceae (Mulberry Family)

A dense, round-headed evergreen tree up to 50 feet high, native to tropical West Africa, this moderately fast growing tree will take sun or shade and has moderate salt, wind, and drought tolerance. It is a good specimen or garden tree, as well as an excellent potted or large tubbed specimen for the lanai or interior. A hedge of Fiddle Leaf Fig makes a large and effective screen or windbreak.

Ficus rubiginosa
RUSTY FIG, PORT JACKSON FIG
Moraceae (Mulberry Family)

An evergreen tree with a broad crown growing to 50 feet high, this useful species from New South Wales, Australia, thrives in heat and is partially drought and salt tolerant and highly wind resistant. An attractive garden tree and proven indoor plant, it may be used in sun in most soils. Selected cultivars are available, some with variegated foliage. Provide adequate space for its wide-spreading roots.

Filicium decipiens
FERN TREE
Sapindaceae (Soapberry Family)

This dense, handsome ornamental tree, up to 50 feet high, is native to India. It is slow growing and should be used in sun or light shade in most good garden soils. It has good wind tolerance and moderate salt tolerance. It may be used as a shade, garden, or street tree. It has nonaggressive roots. Plant it over a ground cover that will absorb the attractive but troublesome small black fruits. It makes an excellent tubbed specimen.

Jacaranda mimosifolia
JACARANDA
Bignoniaceae (Catalpa Family)

This deciduous tree from northwestern Argentina reaches 50 feet in height and bears clusters of showy flowers in spring and summer. Moderately fast growing, it can be used as a street, garden, or shade tree. It is best in sun in fertile, well-drained soils at cooler elevations and has moderate wind tolerance and good drought tolerance. Its tolerance to salt is poor. Its seed pods are used in dried arrangements.

Kigelia africana
SAUSAGE TREE
Bignoniaceae (Catalpa Family)

A round-headed tree up to 50 feet high, this unique species is native to tropical West Africa. Its purple flowers are night blooming. This moderately fast growing tree can be used as a free-standing specimen or shade tree in gardens and parks. Plant it in full sun in most soils. Its tolerance is good to drought, moderate to wind, but poor to salt.

Lagerstromia speciosa
QUEEN'S CRAPE MYRTLE, GIANT CRAPE MYRTLE, PRIDE OF INDIA
Lythraceae (Crape Myrtle Family)

This dense, round-headed tree from India to China and south to New Guinea and Australia produces its clusters of attractive flowers in spring and summer. It is moderately fast growing to 50 feet high and makes a good flowering street, specimen, or garden tree. Performing best in sun in a fertile, moist soil, it has moderate wind tolerance but poor salt tolerance. Dried seed pods are used in arrangements.

Litchi chinensis
LITCHI, LYCHEE
Sapindaceae (Soapberry Family)

A wide-spreading evergreen tree up to 40 feet high, this species from southern China is used as a shade tree in sunny locations in slightly acid, moist soils. Clusters of its edible reddish maroon fruit appearing in late spring and summer add to its landscape attractiveness. It is a fine shade tree but requires protection from harsh winds.

Melia azedarach
PRIDE OF INDIA, PERSIAN LILAC, CHINABERRY (P)
Meliaceae (Mahogany Family)

An open, spreading, semievergreen tree up to 50 feet high, this useful tree is native to Asia. It produces fragrant flowers in the spring and summer followed by small, yellow, spherical fruits that can be used in arrangements. It will tolerate poor alkaline soils and is drought tolerant. Pride of India makes a good shade or park tree and has moderate wind and salt tolerance.

Nageia falcatus
AFRICAN FERN PINE
Podocarpaceae (Podocarpus Family)

An evergreen tree from East Africa growing to 50 feet high, this species may be used in sun or part shade in various moist, well-drained soils. It has good wind and moderate salt tolerance. An excellent specimen or accent tree, it is also useful as a container plant indoors or out. It may be used as a high informal hedge or screen, or, because it tolerates clipping, as a formal hedge, screen, or espalier.

Noronhia emarginata
MADAGASCAR OLIVE (P)
Oleaceae (Olive Family)

Thriving both along windward coasts and inland, this tree reaches 40 feet in height. It is highly wind and salt tolerant and will grow successfully in almost any soil, as well as pure beach sand. Madagascar Olive makes a good shade tree and may be used to form a strong, tall screen or windbreak or readily controlled to make a dense hedge.

Olea europaea subsp. *europaea*
OLIVE, 'OLIWA
Oleaceae (Olive Family)

A bushy evergreen tree up to 35 feet high with soft, gray green foliage, this species is native to the Mediterranean region. It must have full sun, tolerates drought, heat, wind, and poor soils, but does best in rich, deep soils. This slow growing tree can be used as a specimen, screen, or windbreak plant. It is brilliant in the nightscape. It does not bear fruit at low elevations in Hawai'i. Its foliage is the Christian symbol of peace.

Pimenta dioica
ALLSPICE TREE
Myrtaceae (Eucalyptus Family)

From the West Indies and Central America, this evergreen tree grows to 50 feet high. It may be used in sun or shade in any good soil and has good wind, moderate drought, but low salt tolerance. It makes a good shade or specimen tree for garden, street, or park. Its foliage is highly fragrant. Plant Allspice over a ground cover that will absorb the beautiful—but staining—purplish black fruit. The attractive bark of this tree is appreciated in the nightscape. Its dried fruit produces the allspice of commerce.

Pithecellobium dulce 'Variegata'
White 'Opiuma (T)
Fabaceae (Bean Family)

A moderately fast growing, spreading tropical American tree up to 40 feet high, this useful tree is planted for foliage accent, shade, or framing. It is easy to grow in full sun in most well-drained soils and has good drought tolerance and moderate wind and salt tolerance. Prune with care to avoid the sharp spines.

Plumeria obtusa
Singapore Plumeria (P)
Apocynaceae (Dogbane Family)

Native to the West Indies, this plumeria grows to 35 feet high and retains its foliage most of the year. Its large clusters of fragrant flowers appear from spring through fall. This popular plant is used as a specimen tree in the garden or entryway where its shade and fragrance can be appreciated. It is effective used in groupings or as a container plant. Singapore Plumeria does well in hot, semidry locations in sun or light shade in most soils. Flowers are used in arrangements but are not usually used for leis because of their large, rather floppy petals.

'Common Yellow' 'Daisy Wilcox'

'Paul Weissich' 'Thornton Lilac' 'Kāne'ohe Sunburst'

'Kauka Wilder' 'Hilo Beauty'

Plumeria rubra
Plumeria, Temple Tree (P)
Apocynaceae (Dogbane Family)

One of Hawai'i's favorite trees, this tropical American species is prized for its highly fragrant flowers, which are used to make leis. Well over a hundred cultivars have been developed, displaying flower colors ranging from white through light and dark pink, orange, yellow, and gold to red and dark red. Bicolors, semidoubles, and dwarf forms are available. Flower size is variable. Larger flowered cultivars are not generally used for leis. Plumeria grows to 35 feet high in sun or light shade in most well-drained soils, but it flowers best under hot, semidry conditions. It is useful for its shade, as a specimen, or in groupings and has moderate salt and wind tolerance. Plumerias drop their foliage in winter and produce great clusters of flowers just prior to refoliating in the spring, continuing for over six months. Flowering branches are used in arrangements. Selected cultivars include: 'Common Yellow,' a white with a strong yellow center; 'Daisy Wilcox,' a large-flowered pale pink; 'Paul Weissich,' a deep yellow with a gold center; 'Thornton Lilac,' a lavender pink with darker throat; 'Kāne'ohe Sunburst,' a deep pink with yellow throat; 'Kauka Wilder,' red pink with orange center; and 'Hilo Beauty,' a deep red.

Podocarpus elatus
BROWN PINE, PINE-TINT PODOCARPUS
Podocarpaceae (Podocarpus Family)

This Australian evergreen tree may be used in sun or shade in any soil. It has moderate drought and salt tolerance and good wind tolerance. Due to its slow growth rate, this large tree is considered in the medium range for local gardens. It is a good garden or street tree that also can be sheared and used for hedges, screen plantings, and espaliering. It is a good tubbed specimen for the lanai or interior. Foliage may be used in arrangements.

Podocarpus macrophyllus
JAPANESE YEW, KUMASAKI SOUTHERN YEW
Podocarpaceae (Podocarpus Family)

A slow growing, broadly columnar evergreen tree up to 50 feet high, this Japanese species thrives in full sun to deep shade and prefers well-drained, moist soils. It shows good wind tolerance and moderate salt and drought tolerance. It makes a good container specimen both outside and in well-lighted interiors. Japanese Yew makes an informal screen or windbreak and can be clipped for hedge, espalier, or topiary work. Foliage may be used in arrangements.

Prosopis pallida
KIAWE, ALGAROBA, MESQUITE (T)
Fabaceae (Bean Family)

The *Kiawe*, from northern Peru, is an especially useful tree that grows to 50 feet high. It thrives in hot, dry locations in deep soils, but its nitrogen fixing capability makes it tolerant of shallow, rocky soils. It also has excellent drought and salt tolerance and moderate wind tolerance. *Kiawe* provides light shade for the garden and may serve as a partial screen or windbreak planting. A thornless form is available.

Pseudobombax ellipticum
PINK BOMBAX, SHAVING-BRUSH TREE,
PINK SHAVING-BRUSH TREE
Bombaceae (Bombax Family)

Pink Bombax, growing to 40 feet high, is deciduous and has smooth, highly ornamental bark. Spectacular flowers appear in late winter to early spring just prior to the appearance of new mahogany red leaves, which turn green when they mature. There are white and dark pink flowering forms. It thrives in sun or light shade in a fertile, moist, well-drained soil. It has moderate salt, wind, and drought tolerance. Pink Bombax is used for both its flowers and bark as an accent, specimen, or shade tree in the garden and parks. It is native to tropical Mexico.

Medium Trees

Ravenala madagascariensis
TRAVELLER'S TREE, TRAVELLER'S PALM
Strelitziaceae (Bird of Paradise Family)

A banana-like, clumping evergreen tree up to 40 feet high, this unique Madagascan species is used as a spectacular free-standing accent in the larger garden or displayed against a large building. It is best in sun in a rich, well-drained, moist soil. It has only fair salt tolerance and must be sheltered from strong wind to prevent splitting of its large leaves. It is a moderate grower and the enormous inflorescences may be used in arrangements.

Salix babylonica
WEEPING WILLOW, BABYLON WEEPING WILLOW
Salicaceae (Willow Family)

This fast growing, spreading deciduous tree reaches 50 feet in height and is prized for its weeping habit. Probably native to China, it is best used as a single specimen, especially in gardens of Chinese design, or as a bordering grove along a stream or pond side. It requires moisture and is highly tolerant of wet soil conditions. It produces a light shade and is fairly wind tolerant.

Sapindus saponaria
SOAPBERRY, Aʻe
Sapindaceae (Soapberry Family)

Native to a wide range of habitats from tropical Mexico and Hawaiʻi to the South Pacific, *Aʻe* is known elsewhere as Soapberry. Preferring cooler elevations, it also thrives at sea level. It grows to 40 feet high given ample moisture, although it tolerates moderate drought. It is wind resistant. Use *Aʻe* as a foliage accent, for shade, as a light screening tree, and for its small, spherical, jet black seeds, which are made into leis.

Schefflera actinophylla
BRASSAIA, OCTOPUS TREE, UMBRELLA TREE
Araliaceae (Panax Family)

Brassaia is native to tropical Australia. Evergreen, it grows quickly to 40 feet high, producing its octopus-like inflorescences in spring and summer. This unique plant has moderate salt, wind, and drought tolerance and can be used in sun or deep shade in most well-drained soils. It is an excellent accent and a popular container plant either indoors or on the open lanai. It makes a dense screen, windbreak, or sound barrier. Its reddish flower buds are made into leis.

Schinus molle
CALIFORNIA PEPPER TREE, PEPPER TREE (S)
Anacardiaceae (Cashew Family)

This attractive weeping tree comes to us from the Andes of Peru. Evergreen, it reaches 35 feet in height, producing insignificant flowers followed by clusters of small, rose-colored fruits that are used in arrangements. It is tolerant of poor, alkaline soils, has excellent drought tolerance and good wind tolerance, but its tolerance of salt is poor. It is a good specimen tree for light shade in the garden and for avenue and park plantings. It does best in cooler, drier conditions at higher elevations in Hawaiʻi.

Swietenia mahogani
West Indian Mahogany, Cuban Mahogany
Meliaceae (Mahogany Family)

Mahogany, native to southern Florida and the West Indies, is a dense, slow growing evergreen tree reaching 50 feet in height. It grows best in sun but tolerates light shade and is tolerant of many soil conditions. It has good wind tolerance but only moderate salt and drought tolerance. Mahogany is a good shade tree for the larger garden or park and makes an attractive tubbed specimen. Its woody fruit may be used in arrangements.

Syzygium malaccense
Mountain Apple, ʻŌhiʻa ʻAi
Myrtaceae (Eucalyptus Family)

In Hawaiʻi's moist valleys at low elevations, dense groves of ʻŌhiʻa ʻAi thrive and produce their abundant and conspicuous flowers, followed by red—or rarely, white—edible fruits, which are also used in arrangements. Attaining a height of 45 feet, it is a moderately rapid grower. It makes an excellent specimen, background planting, or tall hedge or screen. It is native to India and Malaya.

Tabebuia heterophylla
Pink Tecoma
Bignoniaceae (Catalpa Family)

A slender semievergreen tree reaching 40 feet in height, this species is native to the West Indies. Growth habit and foliage are variable, as is flower color, which ranges from light to dark pink and pinkish purple to white. Flowers appear abundantly in spring but may also be seen occasionally most of the year. It may be grown in sun or partial shade and has moderate drought, wind, and salt tolerance. It is used as a bright flowering specimen or color accent in the garden and in parks and is also useful as an avenue planting.

Tabebuia impetiginosa
Purple Trumpet Tree
Bignoniaceae (Catalpa Family)

A deciduous native of Brazil, this slow growing and spreading tree may eventually reach 50 feet in height. Its large clusters of flowers are produced in the spring, varying in color from white and light pink to purple. This tree can be used as a specimen for the garden or as a street tree. It grows best in sun in a rich, well-drained soil.

Tabebuia rosea
Pink Trumpet Tree, Rosy Trumpet Tree
Bignoniaceae (Catalpa Family)

A showy deciduous tree up to 50 feet high, this species is native from Mexico south to Venezuela and Ecuador and bears large clusters of flowers. Produced periodically during the year, flowers vary in color from dark pink to nearly white. It may be grown in sun or light shade and has moderate wind, salt, and drought tolerance. It is valued for shade and as a spectacular flowering specimen for gardens and parks.

Terminalia catappa
Tropical Almond, False Kamani, Kamani Haole
Combretaceae (Combretum Family)

This spreading, briefly deciduous tree up to 50 feet high is highly tolerant of seaside conditions. It has good wind tolerance and moderate drought tolerance and will thrive with part of its root system exposed to salt water. From the East Indies, it is a good shade tree and foliage color accent for the larger garden or park. Prior to falling, in late autumn or early winter, leaves turn to brilliant golden yellow or red.

Thespesia populnea
Milo, Portia Tree
Malvaceae (Hibiscus Family)

A dense, round-headed evergreen tree reaching 40 feet in height, this species is found on many tropical shores. It is grown in sun or partially shaded locations and is adaptable to a range of soil conditions. *Milo* has excellent salt tolerance. Its tolerance to wind is good and to drought moderate. Use it as a screen or windbreak, a specimen or shade tree for the garden or park, or in avenue plantings. Its fissured bark is attractive. Its yellow flowers are produced much of the year.

Chapter 7
Large Trees

Large trees that grow over 50 feet high are generally best suited to larger properties, parks, school campuses, and other areas where room is available for proper development and display. A few of the more commonly seen large trees are included here.

Araucaria columnaris
Cook Pine
Araucariaceae (Araucaria Family)

A fast growing columnar evergreen tree attaining a height of well over 100 feet, this species is native to the Isle of Pines near the coast of New Caledonia. It is valued as a windbreak, an accent against tall structures, an avenue tree, in parks, or planted in groves. It has moderate salt and drought tolerance and good wind tolerance. It is almost indistinguishable when young from the less commonly seen Norfolk Island Pine (not pictured). Both are used as Christmas trees and for holiday decorating and have similar uses in the landscape. Both make excellent tubbed specimens.

Artocarpus altilis
Breadfruit, 'Ulu
Moraceae (Mulberry Family)

A spreading evergreen tree growing to 60 feet high, Breadfruit is grown for its highly ornamental foliage and for its large green fruits, which are edible and are also used for arrangements. Native to the Malay Peninsula, it has poor wind, salt, and drought tolerance. It is recommended for planting away from paved surfaces and sitting areas where its heavy falling fruit will not become a maintenance problem or liability.

Calophyllum inophyllum
Kamani, Alexandrian Laurel (P)
Clusiaceae (Mangosteen Family)

This slow growing evergreen tree from the shores of the Indian and western Pacific Oceans reaches 60 feet in height. Clusters of fragrant white flowers appear in spring and summer and are used in leis. Its spherical fruit is used both green and dried in arrangements. *Kamani* is an excellent shade tree, planted as a specimen or in a grove. It has been used in parks, school campuses, and avenue plantings. With its dense foliage it makes a good high windbreak or screen. It performs best in sun or light shade, tolerating most garden soils, wind, salt, and drought.

Casuarina equisetifolia
COMMON IRONWOOD
Casuarinaceae (Ironwood Family)

A fast growing spirelike tree up to 80 feet high, this tough species comes from the southern Pacific Islands and regions westward to India. A nitrogen fixer, it is tolerant of most soils, even pure sand. It displays good heat, salt, wind, and drought tolerance. It is used as a windbreak at the seashore, where it is also seen as a clipped hedge. The small spherical fruit may be used decoratively.

Enterolobium cyclocarpum
EARPOD, ELEPHANT'S EAR
Fabaceae (Bean Family)

A fast growing tropical American deciduous tree, the Earpod may ultimately reach 100 feet in height, with an even greater spread. It makes a good shade tree for parks and other larger properties. It grows best in full sun and, being a nitrogen fixer, tolerates most soils. It has excellent drought tolerance but only fair tolerance to wind and salt. The white bark is beautiful in the nightscape. Its fruit is used in arrangements and seeds in jewelry making.

Erythrina variegata
INDIAN CORAL TREE, TIGER'S CLAW, WILIWILI HAOLE (P) (T)
Fabaceae (Bean Family)

This fast growing, spreading, deciduous tree up to 60 feet high bears its large clusters of brilliant flowers in winter before new foliage develops. It makes a fine shade or specimen tree for parks and other large spaces and is used for avenue plantings. It requires full sun and, due to its nitrogen fixing abilities, grows well in almost any well-drained soil. It has good wind tolerance and moderate tolerance to salt and drought. This *wiliwili* is native from India to southern Polynesia. There is a white-flowered and a variegated-leaf form.

Erythrina variegata 'Tropic Coral'
Tropic Coral, Tall Erythrina (P) (T)

A popular, columnar form of *Wiliwili Haole*, this tree is used as a windbreak or screen and near tall buildings when space is limited. Growing to 50 feet high, it displays its bright flowers at the top of the tree in spring, making a strong color accent in the landscape. It retains all of its foliage in full sun and, because it is a nitrogen fixer, tolerates almost any well-drained soil.

Eucalyptus citriodora
LEMON-SCENTED GUM
Myrtaceae (Eucalyptus Family)

A tall, slender evergreen tree from Australia, this species displays clusters of white flowers in winter and bears strongly lemon scented foliage all year. It is fast growing to 75 feet high and produces a light shade, making it suitable as a lawn specimen for parks and planted in groves. It has good drought and moderate wind tolerance but poor salt tolerance. Its attractive bark makes linear accents in the nightscape.

Eucalyptus deglupta
MINDANAO GUM
Myrtaceae (Eucalyptus Family)

This very fast growing evergreen tree from New Guinea to Indonesia and the Philippine Islands ultimately reaches 120 feet in height. Its smooth, colorful bark is its outstanding attraction. Thriving in deep, moist, well-drained soils, this gum makes a fine specimen or grove for large gardens, parks, schools, and broad avenues. It has moderate wind and salt tolerance. Large clusters of white flowers appear in spring.

Ficus benghalensis
INDIAN BANYAN, BANYAN, VADA TREE
Moraceae (Mulberry Family)

This very large evergreen tree up to 80 feet high is broadly spreading via accessory trunks that develop from aerial roots, so it should be limited to parks, golf courses, or other spacious properties. Specimens have been known to spread over several acres. Native to India and Pakistan, it grows in almost any well-drained soil and has moderate drought and salt tolerance and good wind tolerance. It is sacred to Hindus.

Ficus benjamina
BENJAMIN TREE, BENJAMIN FIG, WEEPING FIG
Moraceae (Mulberry Family)

An evergreen tree from India and Malaysia, this fast growing tree reaches 60 feet in height with an even greater spread. It has a dense canopy with a weeping growth habit and highly variable leaves that range from slender to narrowly broad. A highly popular container specimen for the lanai and interior, it makes a good shade or specimen tree for larger gardens or parks. It will tolerate most soils and has moderate wind, salt, and drought tolerance. It has a visually exciting but invasive surface root system, requiring space in the landscape.

Large Trees

Ficus macrophylla
MORETON BAY FIG
Moraceae (Mulberry Family)

An eastern Australian species, this dense evergreen tree reaches 60 feet in height with at least an equal spread. It is fast growing, tolerates most soils, and has good smog, vog, wind, drought, and salt tolerance. Its large leathery leaves and dense canopy make it an ideal screen, high hedge, or windbreak. It is especially effective for noise control. An excellent specimen or shade tree for larger properties and parks, it also works well as a tubbed specimen for the lanai or interior. A variegated form is available.

Ficus microcarpa
CHINESE BANYAN, INDIAN LAUREL, MALAYAN BANYAN
Moraceae (Mulberry Family)

Native to an area from the Malay Peninsula to Borneo, this fast growing evergreen tree up to 60 feet high spreads its dense canopy and produces multiple aerial roots in wet areas. It is tolerant of all soils. It is a massive specimen or shade tree for large properties and parks and has good wind, smog, vog, and drought tolerance as well as moderate salt tolerance. It has a prominent and invasive root system. It makes a good container plant for both exterior and interior places. There is a variegated form available.

Melaleuca quinquenervia
PAPERBARK TREE, CAJEPUT TREE
Myrtaceae (Eucalyptus Family)

The Paperbark Tree, from Southeast Asia and Australia, grows rapidly to 80 feet high. It is distinctive with its broad columnar growth habit and layers of thin, peeling, whitish bark. It has a variety of uses where other plants are difficult to grow. It is tolerant of poor soils, wet or dry soils, fire, heat, wind, salinity and alkaline conditions, and salt winds. Paperbark makes a good windbreak, screen, or enclosure.

Pterocarpus indicus

NARRA, BURMESE ROSEWOOD, BLOODWOOD

Fabaceae (Bean Family)

Narra is a fast growing, stately tree from Southeast Asia and the Philippines that bears masses of bright flowers in spring. Growing to 80 feet high with a somewhat weeping habit, it finds use in parks, as an avenue tree, or as a specimen in a large garden. A nitrogen fixer, Narra grows in almost any well-drained soil and displays moderate wind, salt, and drought tolerance.

Samanea saman

MONKEYPOD, RAIN TREE

Fabaceae (Bean Family)

One of the most popular shade trees in the tropics, this rapid growing deciduous tree from tropical America forms a wide-spreading, umbrella-like crown. It grows to 80 feet high with a greater spread. Attractive flowers appear from late spring to summer. A sun lover, it thrives in all moist, well-drained soils due to its nitrogen fixing ability. It has good drought tolerance and moderate salt and wind tolerance.

Spathodea campanulata

AFRICAN TULIP TREE

Bignoniaceae (Catalpa Family)

A showy, somewhat columnar tree up to 75 feet high, this tropical African species produces its brilliant flowers most of the year. Evergreen, it needs sun with moderate moisture in most soils. It has moderate wind but low salt tolerance. African Tulip is one of the showiest ornamental trees for shade in parks and large gardens. A cultivar with bright orange-yellow flowers is available.

Tabebuia donnell-smithii

GOLD TREE

Bignoniaceae (Catalpa Family)

A handsome, somewhat columnar deciduous tree from Mexico and Guatemala, this showy species reaches 100 feet in height. Flowering season is highly variable, with masses of spectacular blooms appearing either in winter or spring before the new foliage appears; there is occasional flowering later in the year. It is a moderate grower, thriving best in hot, dry areas in almost all well-drained soils. It displays good wind and drought tolerance but low salt tolerance. It is well used in large gardens, parks, groves, and avenue plantings for its shade and as a color accent.

Large Trees

Tabebuia ochracea subsp. *neochrysantha*
YELLOW TRUMPET TREE
Bignoniacae (Catalpa Family)

Frequently flowering several times a year beginning in spring, this spectacular species from Central and South America may reach 60 feet. It has a round-headed canopy and is deciduous, dropping its leaves for a short period in winter. A moderately fast grower, it does well in any well-drained soil, and although tolerating some drought, it grows best when soil is moist. It is fairly wind tolerant but does not like salt air.

Chapter 8
Vines

This grouping of plants includes those that are not self-supporting. They must climb over, cling to, or rest on other plants or structural elements. Some can be used as ground covers or pruned into shrubs.

Allamanda cathartica
COMMON ALLAMANDA (P)
Apocynaceae (Dogbane Family)

This evergreen, ever-blooming vine from tropical America will climb 50 feet. Its large flowers are fragrant and bloom best in full sun in a fertile, well-drained soil. It has moderate drought and wind tolerance but is not tolerant of salts. Use as a bank or deep ground cover, an archway and pergola cover, or prune into a shrub. Several cultivars are available, including the commonly used Henderson Allamanda (*A. cathartica* 'Hendersonii'), which has larger leaves and flowers.

Alyxia oliviformis
MAILE
Apocynaceae (Dogbane Family)

Maile, an endemic Hawaiian vine, is most commonly found in shady, moist areas. There are several leaf sizes and shapes but all are used to make a highly fragrant lei. Use *Maile* to cover a shaded trellis. Give it a rich, well-drained soil for best results. It is important to Hawaiians and is used in religious observances, noted in traditional legends and songs, and associated with the hula.

Antigonon leptopus
MEXICAN CREEPER, CHAIN OF LOVE, MOUNTAIN ROSE
Polygonaceae (Buckwheat Family)

A fast spreading vine from Mexico, this evergreen species climbs to 40 feet by tendrils on a trellis or pergola and may also be used as a rambling ground cover. Its brilliant flowers are borne much of the year. Cultivars with dark pink and white flowers, such as White Coral Vine (*A. leptopus* 'Album'), are available. Use Mexican Creeper in the hottest spot in the garden, in sun, on banks, and on walls. While drought and wind tolerant, it does not stand up to salt air. Flowers are used in arrangements.

Argyreia nervosa
WOOLLY MORNING GLORY, BABY WOOD ROSE (P)
Convolvulaceae (Morning Glory Family)

This twining evergreen vine from India is fast growing, climbing to 40 feet. Its showy flowers are produced from spring to fall. Woolly Morning Glory performs best in sun in a rich, well-drained soil. It displays moderate wind and drought tolerance but is not salt tolerant. It is a useful vine for a trellis or fence. Dried fruit clusters are used in arrangements.

Artabotrys hexapetalus
CLIMBING ILANG-ILANG, YLANG YLANG, LANALANA
Annonaceae (Custard Apple Family)

This woody vine from Sri Lanka and southern India climbs by means of curved hooks on the stem. Heavily scented flowers are produced under the foliage and are followed by grapelike clusters of fragrant yellow fruit. This vine grows best in sun in fertile soil in a protected location. It can be used to cover a pergola, or on slopes, walls, and fences. Flowers can be used in leis.

Asparagus plumosus
ASPARAGUS FERN (T)
Liliaceae (Lily Family)

A woody, climbing evergreen vine from South Africa with stems up to 20 feet long, this species grows best in partial shade in most well-drained soils. It is used to cover fences and trellises. Several cultivars are available. The fernlike foliage is frequently used in corsages and flower arrangements.

Bauhinia galpinii
RED BAUHINIA, NASTURTIUM BAUHINIA
Fabacaeae (Bean Family)

This evergreen plant from tropical Africa climbs to 40 feet but is readily kept as a wide-spreading shrub up to 15 feet high. Its showy flowers are produced from spring to fall. Flower color varies from brick red to strong yellow red. Red Bauhinia grows best in full sun in a moderately fertile, well-drained soil. It shows moderate drought and wind tolerance but is not tolerant of salts. It is a useful vine for a trellis or pergola or as a spreading bank cover and can be used as a large specimen plant or informal hedge.

Beaumontia multiflora
EASTER LILY VINE
Apocynaceae (Dogbane Family)

A slow growing, woody, twining vine from India, this species climbs to 25 feet and is used to cover fences or pergolas. Large flowers are produced at branch tips and have a spicy fragrance. The Easter Lily Vine grows best in full sun in well-drained soils containing organic matter. It will tolerate light shade and is moderately drought tolerant.

Bougainvillea spectabilis
BRAZIL BOUGAINVILLEA (T)+
Nyctaginaceae (Four O'Clock Family)

A large vine reaching 50 feet in length, this Brazilian native produces masses of bracts ranging from purple to rose to dusty red. Requiring full sun for best bloom production, it will thrive in almost any well-drained soil with moderate moisture. Bougainvillea has moderate drought and salt tolerance. It is prized for its ability to cover large pergolas and banks in large gardens or parks. However, since it is easily controlled by pruning, most gardeners find Bougainvillea a highly desirable color accent even in small landscapes. It also makes a fine container specimen. Flower production is enhanced by pruning. Dozens of cultivars are available, with bract colors ranging from pure white through pink, orange, gold, red, and lavender, as well as a bicolor. Several have double bracts and foliage variegated with white or yellow. Among the most popular are: *B.* 'California Gold'; *B.* 'Carmencita,' with double, reddish purple bracts; *B.* 'Mary Palmer,' a bicolor displaying both white and rose pink bracts, as well as showing both colors on the same bract; *B.* 'Miss Manila,' possibly the most popular of the cultivars, almost ever-blooming, with bracts showing both orange and violet; and *B.* 'Raspberry Ice,' which is one of the variegated cultivars.

'California Gold'

'Carmencita'

'Mary Palmer'

'Miss Manila'

'Raspberry Ice'

Cissus discolor
TRAILING BEGONIA, REX BEGONIA VINE
Vitaceae (Grape Family)

Native to Indonesia, this colorful evergreen species climbs by means of tendrils to 15 feet. Its inconspicuous flowers are clustered in the leaf axils. Trailing Begonia is best in a shaded location in a moist, well-drained soil. It makes an attractive potted or hanging basket plant for the interior or lanai, a colorful foliage specimen in the garden, and a manageable vine on trees and fences.

Clerodendrum x *speciosum*
CLERODENDRUM VINE
Verbenaceae (Verbena Family)

An evergreen hybrid vine climbing to 30 feet and producing showy flowers in winter and spring, this highly ornamental plant will cover a trellis, arbor, or fence. The persistent calyx is used in leis. Plant in good, well-drained soil in a sunny place. It is a fairly rapid grower.

Clerodendrum splendens
RED CLERODENDRUM
Verbenaceae (Verbena Family)

Climbing rapidly, this tropical African vine reaches 30 feet in length. Its large clusters of brilliant flowers appear abundantly during winter. This species is best used in full sun on vertical supports such as a fence or trellis, and it can be trained along eaves given support. It will grow in most garden soils and needs protection from strong winds. It is neither drought nor salt tolerant.

Clerodendrum thomsonae
BLEEDING HEART VINE
Verbenaceae (Verbena Family)

A twining, shrubby evergreen vine from tropical West Africa, this species reaches 20 feet in length and produces its clusters of flowers in late summer and fall. It performs best in full sun in a rich, loose, moist soil. It makes an excellent trellis or fence cover, a potted specimen, or is readily maintained as a low border or hedge.

Clitoria ternatea
BUTTERFLY PEA, BLUE BUTTERFLY PEA (P)
Fabaceae (Bean Family)

A soft, slender, twining vine reaching 15 feet in length, this species, probably originating in tropical Asia, displays its startling flowers in summer. Double- and white-flowering forms are known. Fast growing, Butterfly Pea may be grown in sun or part shade in most soils. It provides quick cover for a fence or trellis or may be used as a potted plant on a sunny deck or lanai.

Congea griffithiana
SHOWER OF ORCHIDS VINE, CONGEA, PINK SANDPAPER VINE
Verbenaceae (Verbena Family)

This sprawling plant from Malaysia may be used as a vine or readily trained as a shrub. Abundant sprays of flowers are produced during late winter and spring. It is used to clamber over a fence or arbor or as a bank cover. It is a good color accent, either as a specimen plant or in a mass planting for large spaces. Growing best in full sun to 20 feet, it is tolerant of a wide range of soils. It has moderate tolerance to drought but not to salt.

Cryptostegia madagascariensis
MADAGASCAR RUBBER VINE (P)
Aesclepiadaceae (Milkweed Family)

This rapid growing, twining woody vine from Madagascar may climb 60 feet, providing excellent cover for fences and arbors. Clustered flowers appear in winter and spring. Removal of seed pods prior to maturing is recommended as it reseeds readily and can become invasive. Madagascar Rubber Vine is exceptionally drought and wind tolerant and moderately salt tolerant. It is a good choice for the Xeriscape.

Epipremnum pinnatum 'Aureum'
GOLDEN POTHOS, TARO VINE (A)
Araceae (Aroid Family)

A sturdy vine from the Solomon Islands, this species may climb 40 feet, attaching its variegated stems to almost any support by means of special roots. Juvenile leaves are small and heart shaped, marbled with yellow. As the vine climbs, leaves become greatly elongated and broadened, carrying the variegations more as irregular splashes. There is a form with unvariegated foliage. All are used to cover concrete or stone walls or permitted to clothe tree trunks. It is not recommended for wooden fences as its roots will destroy the wood. Plants with juvenile foliage make excellent shaded ground cover and ideal house plants. It requires shade and moisture. There are other cultivars available.

Ficus pumila
CREEPING FIG, CLIMBING FIG
Moraceae (Mulberry Family)

A vigorous evergreen vine from south China through Malaysia, this species may climb 100 feet, aggressively attaching itself to almost any surface by special roots. Leaves on vegetative growth are small, while fruiting stems bear larger leaves. It is used to cover concrete and rock walls, even sides of buildings, road cuts, and bridge abutments. Several cultivars are available with dwarf and variegated foliage. These are used as potted specimens both for interior and lanai use. Creeping Fig grows easily in almost any soil, in sun to deep shade, and prefers moisture but tolerates moderate drought.

Ipomoea horsfalliae
KŪHIŌ VINE, PRINCE'S VINE
Convolvulaceae (Morning Glory Family)

A fast growing, twining perennial from the West Indies, this vine climbs to 30 feet. The evergreen Kūhiō Vine displays its showy clusters of flowers from fall through spring. It may be used in high or medium light in most soils to cover fences and pergolas or to cascade over a wall or from a hanging planter.

Jasminum laurifolium forma *nitidum*
CONFEDERATE JASMINE, ANGEL WING JASMINE
Oleaceae (Olive Family)

Native to India, this moderately fast growing, climbing species can be used as a vine growing to 20 feet or easily kept as a shrub with proper pruning. Its fragrant flowers are produced much of the year. Confederate Jasmine may be used in foundation groupings, as a ground cover, in planters and shrubbery borders, or as a container plant. It can be used in full sun or broken shade and will tolerate many soil conditions. It has moderate drought and salt tolerance.

Jasminum multiflorum
STAR JASMINE
Oleaceae (Olive Family)

This densely pubescent climber or spreading shrub mounds from 4 to 6 feet in the open but reaches 20 to 30 feet with support. An evergreen plant from India, it produces clusters of flowers much of the year. Star Jasmine is best in sun in a well-drained, fertile soil, but is adaptable to various conditions. It may be used as a vine to cover a fence, pergola, or trellis, but it may also be effective in mass plantings, as a bank cover, or as a low hedge.

Lonicera × *heckrottii*
GOLDEN FLAME HONEYSUCKLE, PINK HONEYSUCKLE
Caprifoliaceae (Honeysuckle Family)

An evergreen, twining, vinelike plant growing to 15 feet, this hybrid bears its colorful flower clusters during spring through summer. Use this attractive plant in sun or shade in most soils as a ground or bank cover, on a fence or trellis, or train as an espalier. It has moderate drought and salt tolerance and excellent tolerance to wind.

'Halliana'

Lonicera japonica
JAPANESE HONEYSUCKLE
Caprifoliaceae (Honeysuckle Family)

A twining evergreen vine from eastern Asia with stems up to 30 feet long, this aggressive vine makes a good fence or arbor cover or may be used as a low ground cover. Its fragrant flowers are seen much of the year and are used in lei making. Grow honeysuckle in sun or light shade in moist, well-drained soils. Available cultivars include Hall's Japanese Honeysuckle (*L. japonica* 'Halliana'), a more vigorous plant, and Purple Japanese Honeysuckle (*L. japonica* 'Atropurpurea').

'Atropurpurea'

Macfadyena unguis-cati
CAT'S-CLAW CREEPER
Bignoniaceae (Catalpa Family)

A vigorous, woody tropical American vine with stems up to 60 feet long, this evergreen species clings to almost any surface by means of clawlike tendrils. Spring flowering, its bright flowers may blanket a fence or wall and hang curtainlike from a tree. Plant it in full sun or light shade. It will tolerate most soils and is quite drought tolerant.

Mandevilla × *amabilis*
MANDEVILLA, DIPLADENIA
Apocynaceae (Dogbane Family)

A moderately fast growing, twining evergreen climber from southeastern Brazil, this species grows to 30 feet long. Its clusters of bright flowers appear from spring to fall. Prized as a potted specimen for the lanai, it is also used to cover trellises and fences. Flowering is heavier when Mandevilla is planted in a rich, moist, well-drained soil in full sun. Several cultivars are available, including Mandevilla Alice Dupont (*M.* × *amabilis* 'Alice Dupont'), with larger, darker pink flowers.

Mansoa hymenaea

GARLIC VINE

Bignoniaceae (Catalpa Family)

A moderately fast growing South American evergreen vine climbing by means of tendrils, this species reaches 40 feet in length. Its foliage is strongly garlic scented—an attraction for some, a drawback for others. Clusters of flowers are produced periodically throughout the year. Growing best in full sun, it does well in almost any soil, including poor, sandy soils. It requires little care and may be used as an arbor or fence cover.

Marsdenia floribunda

STEPHANOTIS, MADAGASCAR JASMINE

Asclepiadaceae (Milk Weed Family)

Native to Madagascar, the popular Stephanotis is a moderately fast growing, twining evergreen vine reaching 30 feet in length. Prized for its clusters of highly fragrant flowers displayed in late winter and spring, it is used in arrangements, bouquets, and leis. In the garden use Stephanotis to cover a fence or trellis or to train along eaves on a support. Flower production is best in full sun. It thrives in most well-drained soils, is heat tolerant, and has moderate drought and salt air tolerance.

Merremia tuberosa

WOOD ROSE, CEYLON WOOD ROSE, PILIKAI (P)

Convolvulaceae (Morning Glory Family)

This is a very vigorous, woody twiner from South America growing to 100 feet. Its attractive flowers are borne in late winter and spring and are followed by brown, rounded fruiting capsules prized for use in dry arrangements. Wood Rose thrives in almost any moderately moist soil in full sun. It is partially drought tolerant. It is good for quick cover but requires large support structures and may easily get out of bounds.

Monstera deliciosa

MONSTERA, CERIMAN, SWISS CHEESE PLANT (P) (A)

Araceae (Aroid Family)

Central America is the origin of this popular ornamental vine. Evergreen, it climbs 30 feet on a concrete or stone wall or tree trunk by means of special roots along its husky stems. It is not recommended for covering wood surfaces, which will be destroyed by the attaching roots. Juvenile leaves are entire. As the vine climbs, mature leaves develop that are deeply cut and perforated. These are used in arrangements. Monstera makes an excellent potted or tubbed specimen both indoors and on the shaded lanai. It can also be used as an effective, tall ground cover. Grow it in a rich, moist soil. Variegated cultivars are available.

Mucuna novoguineensis

RED JADE VINE, SCARLET JADE VINE

Fabaceae (Bean Family)

A woody species from New Guinea climbing 75 feet, this vine is ideal for covering a large pergola or arbor, which perfectly displays its brilliant, neonlike, pendant inflorescenses. It is also trained along eaves, given a strong support. Flowering in spring, blossoms can be used in arrangements. Plant this spectacular vine in a rich, well-drained, moist soil in full sun. It is not tolerant of wind, drought, or salt.

Passiflora edulis

PASSION FRUIT, LILIKOʻI

Passifloraceae (Passion Flower Family)

A vigorous vine from Brazil climbing rapidly to 50 feet by means of tendrils, this evergreen species produces unique flowers throughout much of the year. It works well as a fence cover, on an arbor, or as a large espalier. It has moderate drought and salt tolerance and total wind tolerance. The edible fruit is also used in arrangements. Thriving in almost any well-drained soil, its best response is in full sun.

Passiflora vitifolia

RED PASSION FLOWER

Passifloraceae (Passion Flower Family)

This fast growing evergreen vine from Venezuela to Bolivia is a free-blooming plant with showy flowers produced throughout the year. It climbs by tendrils and may be used to cover a fence, trellis, or pergola. The Red Passion Flower is best in full sun and is tolerant of most soils. It has low salt tolerance and moderate tolerance to drought and wind.

Pentalinon lutea

YELLOW MANDEVILLA

Apocynaceae (Dogbane Family)

This twining evergreen vine from south Florida and the West Indies produces clusters of showy, bright flowers much of the year. It has a moderate growth rate and may be easily espaliered on a wall or fence or used on a pergola, trellis, or lanai post. Yellow Mandevilla is best in full sun in a loose, well-drained soil but is tolerant of some shade and most soil conditions. It tolerates wet conditions and has good drought and salt tolerance.

Petrea volubilis

SANDPAPER VINE, PURPLE WREATH

Verbenaceae (Verbena Family)

Native to the West Indies, Mexico, and Central America, this colorful vine will climb 35 feet. Its name is derived from the sandpapery surface of its foliage. The unusual blooms produced at branch tips appear in winter and spring. There is a white-flowered form available. Used as cover for an arbor, it is also effective as a large semishrubby bank cover. It flowers best when grown in full sun, has some drought and salt tolerance, and is highly wind resistant. Flowers may be incorporated into leis.

Philodendron scandens forma *micans*

VELVET LEAF PHILODENDRON, HEART LEAF PHILODENDRON (P) (S)

Araceae (Aroid Family)

An evergreen vine from Panama climbing to 50 feet, this species produces small, velvety, heart-shaped juvenile leaves and much larger, smooth mature leaves. The juvenile form is a popular interior plant with its tolerance of low light and is excellent in a hanging basket, shaded window box, or terrarium. It requires shade, moisture, and well-drained soils. In the garden it can be used to cover a tree trunk or as a specimen on the lanai. It may also be used in arrangements.

Podranea ricasoliana

PINK TRUMPET VINE

Bignoniaceae (Catalpa Family)

A climbing or sprawling shrub from South Africa, the Pink Trumpet Vine produces showy flower clusters much of the year. It is best in sun in a rich, well-drained soil and has good wind, drought, and salt tolerance. Climbing to 20 feet, it is used on arbors or as a bank cover; it can also be trained as a specimen or screening plant.

Poranopsis paniculata

SNOW CREEPER, PORANA, CHRISTMAS VINE

Convolvulaceae (Morning Glory Family)

A fast-climbing, twining vine up to 60 feet long, this evergreen species from India to Myanmar is used to cover large arbors or as a massive bank cover. It can be utilized on wire mesh along eaves. Planted in sun in a good, well-drained soil, it provides impressive flowering from winter to spring.

Pseudogynoxys chenopodioides

MEXICAN FLAME VINE, ORANGE-FLOWERED SENECIO (P)

Asteraceae (Sunflower Family)

A trailing shrub or smooth, twining vine growing 40 feet in length, this Mexican evergreen plant displays its brilliant clusters of flowers from spring to fall. It grows easily in sun or partial shade in most soils. It has good drought and wind tolerance and moderate salt tolerance. It is used to cover fences and pergolas and, if unsupported, makes a good ground or bank cover.

Pyrostegia venusta

ORANGE TRUMPET VINE, FIRECRACKER VINE, HUAPALA

Bignoniaceae (Catalpa Family)

Native to Brazil and Paraguay, this rapid growing evergreen vine climbs 40 feet by means of strong tendrils. Abundantly flowering in winter and spring, it is used to cover fences and pergolas, as a bank cover, and to tumble over the top of retaining walls. It grows best in full sun, tolerating any soil with moderate moisture.

Quisqualis indica

RANGOON CREEPER

Combretaceae (Combretum Family)

A woody climbing shrub from Myanmar to New Guinea, Rangoon Creeper grows to 30 feet, bearing its clusters of fragrant flowers most of the year. Use it for covering a large vertical trellis, as an espalier, or to cover an arbor. It tolerates a wide range of soils and is moderately tolerant to wind and drought but not salt.

Saritaea magnifica
PURPLE BIGNONIA
Bignoniaceae (Catalpa Family)

A climbing woody vine from Colombia, this species climbs to 60 feet. Its attractive clustered flowers are produced in the fall. It makes an excellent fence cover or could be trained onto a large arbor for additional shade in the garden. It thrives in most soils in full sun and has moderate wind and drought tolerance but no salt tolerance.

Solandra maxima
CUP OF GOLD, CHALICE VINE, IPU KULA (P)
Solanaceae (Potato Family)

A smooth, woody evergreen vine from Mexico, Cup of Gold produces large, fragrant flowers much of the year but more abundantly in winter and spring. Flowers open in the evening. With stems 40 feet long, it makes an ideal arbor cover or can be used on a large vertical trellis. It must be fastened to its support initially as it lacks tendrils or a twining habit. Cup of Gold is also used as a massive bank cover in moist areas. It does best in full sun but tolerates light shade.

Stigmaphyllon floribundum
ORCHID VINE
Malpighiaceae (Malpighia Family)

A twining, woody, tropical American vine, growing to 30 feet in length, this evergreen bears clusters of flowers from spring to fall. It does best planted in a rich soil with ample moisture in sun or partial shade. It has moderate salt air and drought tolerance. It is useful as a fence, arbor, or trellis cover.

Strongylodon macrobotrys
JADE VINE
Fabaceae (Bean Family)

Native to the Philippines, this spectacular, twining evergreen vine rapidly reaches 60 feet in length. It is prized for its long racemes of large flowers that appear abundantly in spring. The flowers are used in arrangements and in leis, although care must be taken by the wearer as the blossoms stain. The Jade Vine thrives when planted in a rich, moist, well-drained soil in full sun. It is an excellent cover for a large pergola, where its pendant flowers are best displayed. It can also be trained along eaves if given strong support.

Syngonium auritum
FIVE FINGERS SYNGONIUM
Araceae (Aroid Family)

A stout evergreen vine from Jamaica, this aroid climbs to 30 feet by means of aerial roots. Its highly ornamental foliage makes it an attractive cement or stone wall cover. It is not recommended for use on wood structures, which can be damaged by its clinging roots. Without support, this vine makes an excellent ground cover. For best results it requires shade, moisture, and protection from strong winds. It thrives in almost any soil.

Syngonium podophyllum
NEPHTHYTIS
Araceae (Aroid Family)

An evergreen vine from Mexico to Panama climbing to 30 feet by means of aerial roots that cling to most surfaces, Nephthytis is a good cement or stone wall cover. It is not recommended for use on wooden structures, which may be damaged by its roots. Juvenile leaves are narrow, becoming divided as the vine climbs and matures. Many cultivars are available with white or pink variegated leaves in the juvenile form. All are popular potted indoor plants or garden ground covers. One of the most useful is White Butterfly Nephthytis (*S. podophyllum* 'White Butterfly').

Tecomanthe dendrophila
TECOMANTHE
Bignoniaceae (Catalpa Family)

A large, twining deciduous vine from New Guinea, Tecomanthe rapidly climbs to 50 feet, producing large clusters of flowers directly off its main stems in winter and early spring. It requires a large pergola or arbor to accommodate its vigorous growth. A large vertical trellis better displays its flowers. Plant Tecomanthe in a rich, well-watered and well-drained soil in full sun.

Thunbergia grandiflora
BENGAL TRUMPET VINE, BLUE TRUMPET VINE
Acanthaceae (Acanth Family)

A vigorous and twining Indian evergreen species, this vine climbs to 80 feet and produces long, pendant racemes of its attractive flowers much of the year. It is commonly used to cover large arbors, where its flowers are seen to best advantage; it can also be trained along eaves if given strong support. It grows well in sun or partial shade in most garden soils. Cultivars include the White Bengal Trumpet Vine (*T. grandiflora* 'Alba'), also with attractive, pendant clusters of flowers. Its landscape uses are the same as the Bengal Trumpet Vine.

Thunbergia laurifolia
LAUREL-LEAVED THUNBERGIA
Acanthaceae (Acanth Family)

An evergreen Indian twining vine climbing to 80 feet, this species bears long, pendant, highly attractive inflorescences much of the year. Planted in full sun in a rich, moist, well-drained soil, this vine makes a rapidly growing cover for a large arbor, where the flowers show off to best advantage. It can be trained along eaves if given strong support.

Thunbergia mysorensis

MYSORE TRUMPET VINE

Acanthaceae (Acanth Family)

Native to India, this twining evergreen vine displays showy, long, pendant inflorescences much of the year, more heavily in spring. It makes an ideal arbor cover, where the flowers can be seen to best advantage, and it can also be trained along eaves if given adequate support. It grows to 35 feet long. It may be used in sun or shade and prefers a well-drained, loamy soil.

Trachelospermum jasminoides

CONFEDERATE JASMINE, STAR JASMINE, MAILE HAOLE

Apocynaceae (Dogbane Family)

This is an evergreen vine from China that climbs to 30 feet, carrying small clusters of highly fragrant flowers in spring and summer. It will cover a trellis or arbor with its strong, twining stems or, without support, makes a dense ground cover. It can be used as a tubbed specimen on a lanai or deck. Plant it in sun or partial shade in most garden soils. It performs better at cool elevations. Confederate Jasmine has moderate drought and salt air tolerence. Several cultivars are available.

Tristellateia australasiae

BAGNIT VINE, CLIMBING SPRAY OF GOLD

Malpighiaceae (Malpighia Family)

A woody evergreen twiner up to 20 feet long, this moderate grower is native to an area from Southeast Asia to New Caledonia. It produces many attractive clusters of flowers in terminal racemes much of the year. It is used as a fence, trellis, or arbor cover or, without support, as a bank or ground cover. It is best in sun in most soils and displays moderate wind, salt air, and drought tolerance.

Chapter 9
Bromeliads, Ferns, and Marantas

Bromeliads

There are some forty-six genera in the Bromeliad (Bromel) or Pineapple Family (Bromeliaceae) that carry about 2,100 species. All but one are found in the New World, and they vary in size from tiny plants an inch or so across to giants whose rosetted foliage spreads 10 feet or more. Bromeliad habitats range from the southern United States to the tip of South America, from sea level to 14,000 feet, and from cold places and hot deserts to tropical rain forests, in sun and in shade; many species are terrestrial, while others are epiphytic. Of the many species and hybrids locally available, we list only a few—those we feel cover the range of sizes, shapes, and colors most useful in the garden. Current taxonomic study may soon modify the Latin binomials of some of these species. Most bromeliads may be used in arrangements.

Aechmea blanchetiana
ORANGE BLANCHETIANA, LEMON BLANCHETIANA (T)

This Brazilian epiphyte grows readily to 3 feet high in a very porous soil. It requires full sun to bring out its strong colors. It tolerates heat, some drought, and seaside conditions. There are two leaf color forms: orange and yellow. Use this species as a strong color accent and as a potted specimen.

Aechmea fasciata
URN PLANT, SILVER VASE (T)

This epiphyte from southern Brazil displays 18-inch leaves that form a vaselike shape. The colorful inflorescence is long lasting. Plant in light shade and keep well watered. Use this plant as a potted specimen or in a porous medium as a ground cover or border. There are several varieties with reddish purple leaves and with green, white, and yellow striping.

Ananas comosus var. *variegatus*
VARIEGATED PINEAPPLE (T)

This is a smaller, variegated form of the commercial pineapple, showing leaves striped longitudinally in whitish cream and green. The entire plant, including its small fruit, becomes pinkish when grown in full sun. It makes an excellent potted specimen for the sunny lanai or as a bright accent in the landscape, tolerating heat, wind, moderate drought, and salt air exposure.

89

Cryptanthus 'Ti'

CRYPTANTHUS TI

One of a great number of highly colorful cryptanthus hybrids available, this small terrestrial bromeliad must have shade, an open, porous soil high in organic matter, and regular, ample watering. It is an excellent ground cover and can be used as a potted specimen for a well-lighted interior space. Cryptanthus cultivars range in color from chartreuse to bright green, pink, and orange to bronze. Leaves may be variously striped or banded with chocolate or silver.

Cryptanthus zonatus

ZEBRA PLANT

Native to Brazilian rain forests, this striking species, a parent of many of the banded hybrids, requires shade and moisture. The Zebra Plant is a popular house plant but finds its highest use in the landscape as a ground or bank cover.

Neoregelia carolinae forma *tricolor*

STRIPED BLUSHING BROMELIAD (T)

This Brazilian bromeliad provides striking color in the landscape, forming a rosette of 12-inch leaves. It is best in partial shade in moist soil conditions. Mature plants turn pinkish at the onset of flowering, the heart of the rosette of leaves becoming vivid red and lasting most of the year. Use it as a potted specimen or ground cover or border.

Neoregelia olens '696'

696

Forming a rosette only 6 inches across, this Brazilian bromeliad spreads by runners that form a mat of strong color in full sun. It is useful in a hanging basket, potted, or as a ground cover in a well-drained open mulch. It does well as an epiphyte, forming a colorful mass of branch-hugging plants as well as hanging clusters.

Tillandsia cyanea

KAMEHAMEHA'S PADDLE, PINK QUILL

This Ecuadorean epiphyte grows in full sun or light shade. It can be used as a ground cover or container plant in a highly porous soil or used to enhance groupings of other epiphytes. Emerging from the rosette of dark, 12-inch green leaves is a long-lasting bright pink and blue inflorescence.

Tillandsia rothii
ROTH'S TILLANDSIA

This slow growing, partially drought tolerant epiphyte from the tropical west coast of Mexico forms a rosette of leaves frequently 18 inches across. It will grow in shade but produces maximum color in strong light. The long-lasting inflorescence displays a pale green flower emerging from a cherry, green, and yellow mass of bracts. Do not let water stand in leaf axils. Use this bromeliad as a color accent among other epiphytes or as a potted specimen.

Tillandsia tectorum
ROOFTOP TILLANDSIA

An easily grown epiphyte from Ecuador and Peru, this species requires bright light and controlled watering to prevent constant wetness. Growing to 1 foot high and 8 inches across, its silver color provides a glowing accent in the landscape and nightscape. Its flowers are purple and violet emerging from a pinkish mass of bracts.

Tillandsia xerographica
XEROGRAPHICA

Deserts of Mexico, Guatemala, and El Salvador are the native habitat of this epiphyte. Overwatering must be avoided. Its silvery white rosette of leaves will reach 1 foot across. The rosette is topped by a chalky inflorescence that is multicolored in salmon to clear pink, yellow, and mauve. Potted or used as an epiphyte in full sun, this bromeliad provides a shining accent in both landscape and nightscape.

Vriesia gigantea seidelii 'Nova'
NOVA

Forming a rosette of lightly banded silver green foliage 4 feet across, this Brazilian bromeliad prefers light shade in a moist, porous medium. In the landscape it provides a striking accent among smaller, darker species and grows well as a potted specimen. It is a glowing accent in the garden at night.

Vriesia hieroglyphica
KING OF BROMELIADS

From southern Brazil, this bromeliad carries 5-foot long rosetted leaves that are shiny, bright green, and spineless, banded with darker green or purplish black, and topped by a tall, yellow inflorescence. It requires shade and moist soil for best results. It provides a strong accent or ground cover in the landscape or makes an excellent potted specimen for the shady lanai or terrace.

Bromeliads, Ferns, and Marantas

Vriesia imperialis
IMPERIAL VRIESIA

This large species from Brazil bears spineless leaves up to 5 feet long. Full sun intensifies the color. It is easily grown in well-drained soil or epiphytically and tolerates drought, wind, and moderate exposure to salt air. It also grows well in high rainfall areas. This plant provides a startling accent in the landscape and makes an excellent large containerized specimen.

Vriesia aff. *regina*
SILVERY VRIESIA

This is a silvery green-leaved, spineless species from Brazil with leaves up to 4 feet long. Leaf color is best in strong light but develops well in light shade. It prefers a cool, moist situation and a rich soil. Use it as a bright accent in the landscape, in a rockery or as a potted specimen for the sunny lanai.

Ferns

Ferns are among the gardener's favorite plants. They are organized into a great number of families covering at least 12,000 species. Fern taxonomy is in a great state of flux; experts do not agree.

Ferns inhabit a wide range of environments and come in a wide variety of sizes and shapes. Most come from moist, warm areas and moist, cool environments, but there are also those adapted to hot, dry places. The following selections—a very small slice of locally available fern species—are those thriving in warm, moist situations and represent a range of landscape uses.

Adiantum capillus-veneris
SOUTHERN MAIDENHAIR FERN, VENUS HAIR FERN, 'IWA 'IWA
Pteridaceae (Pteris Fern Family)

Native to the tropics worldwide, including Hawai'i, Maidenhair Fern is a favorite for the lightly shaded garden as a ground cover and in containers or hanging baskets. Thriving also indoors in a well-lighted space, it reaches 18 inches in height when given ample moisture and regular feeding. Many named cultivars are available in the trade.

Angiopteris evecta
MULE'S FOOT FERN, GIANT FERN
Marattiaceae (Marattia Family)

This is truly a giant among ferns, with fronds reaching a length of 15 feet or more. It requires light shade, regular feeding, and ample moisture. It provides a major accent in the fernery or general garden and is spectacular when night lighted. It is native to an area from Malaysia through southern Polynesia.

Asplenium nidus
BIRD'S-NEST FERN, 'ĒKAHA, 'ĀKAHA
Aspleniaceae (Spleenwort Family)

Native to a broad area from Hawai'i through Polynesia, Australia, Malaysia, and Madagascar, Bird's-Nest Fern is a favorite garden accent for the shaded garden. It can be grown both epiphytically and in the ground. It makes an excellent tubbed specimen. With ample watering and feeding and protection from strong winds, its leaves can grow to 6 feet long, producing a large rosette of foliage whose translucence is dramatized by lighting from below in the nightscape. The leaves are used in arrangements.

Cibotium glaucum
HAWAIIAN TREE FERN, HĀPU'U
Dicksoniaceae (Dicksonia Family)

Endemic to Hawai'i, this is one of the most common of the native tree ferns available for garden use. It is very slow growing, developing a trunk up to 15 feet high crowned with arching fronds of lacy foliage. It may be used in sun or light shade but always in a moist, well-drained location. It has some salt tolerance, but needs protection from hot sun and drying winds in sunny, lowland locations. *Hāpu'u* makes an excellent canopy for shade-loving plants or a specimen either in the landscape or containerized. It is excellent in the nightscape.

Cyathea cooperi
AUSTRALIAN TREE FERN
Cyatheaceae (Tree Fern Family)

In moist areas in rich soil, this tree fern grows rapidly to 15 feet high or more. Fronds may attain a length of 8 feet. Showing more vigor in cooler areas, it forms a delicate canopy for shade lovers. It makes a strong textural accent. The Australian Tree Fern adapts readily to tub culture as long as watering is heavy and consistent. It will thrive in full sun but benefits from noon shading along dry leeward coasts. The fern is spectacular with night lighting from below. Cool rain forest areas of Queensland are its native home.

Davallia fejeensis
LACY HARE'S-FOOT FERN
Davalliaceae (Davallia Family)

Growing to 2 feet high, this Fijian fern forms an excellent ground cover in a shady place in the garden or may be grown epiphytically if given ample water. It does well in hanging baskets where its lacy foliage can be shown to advantage. This fern is particulary beautiful with subtle lighting in the nightscape. The foliage is used in arrangements.

Microlepia strigosa
PALAI, PALAPALAI
Dennstaediaceae (Hay-Scented Fern Family)

This fern is native to the area from Japan westward to India and eastward through Polynesia to Hawai'i. Growing to 2 feet high, it forms an excellent ground cover in partially shaded, wind protected, moist areas and is a useful potted or tubbed specimen. Its delicate fronds are traditionally used in making leis and it was one of the plants placed on the altar of Laka, the hula goddess. It is also used in arrangements.

Microsorum punctatum 'Cristatum'
CRESTED FERN
Polypodiaceae (Sword Fern Family)

From a wide area ranging from tropical Africa to French Polynesia, this 2-foot fern makes a good ground cover in a partly shaded, moist situation. It is a good potted or tubbed specimen. The bold, somewhat stiff fronds are held vertically and are used in arrangements.

Nephrolepis cordifolia
NARROW SWORD FERN, FISH BONE FERN, KUPUKUPU LAU LI‘I
Nephrolepidiaceae (Sword Fern Family)

A native and pantropical fern, the Narrow Sword Fern is somewhat variable as to frond length. Selected forms may exceed 2 feet in length and tend to grow rather vertically. It is tolerant of full sun and partially drought tolerant but responds best in light shade with ample watering. It forms a dense ground cover. Fronds are also used in making leis and in arrangements.

'Bostoniensis'

'Dallas'

'Fluffy Duffy'

Nephrolepis exaltata
COMMON SWORD FERN, KUPUKUPU, NI‘ANI‘AU
Nephrolepidiaceae (Sword Fern Family)

Native and pantropical in distribution, the Common Sword Fern forms a lush ground cover up to 3 feet high in moist, shaded areas. The fronds are used in arrangements and in leis. It can also be planted in hanging baskets where, with regular and ample watering, it makes an excellent display. Among the many popular cultivars, the Boston Fern ('Bostoniensis') is commonly seen, while the Dallas Fern ('Dallas') is popular as a low ground cover up to 12 inches high. Of special note is the unique Fluffy Duffy ('Fluffy Duffy'), prized as a house plant.

Nephrolepis falcata 'Furcans'
FISHTAIL FERN, HI‘UI‘A
Nephrolepidiaceae (Sword Fern Family)

Probably originating in Melanesia, this useful fern is now widespread in the tropics and has naturalized in Hawai‘i. Arching fronds may grow as long as 3 feet. The plants form dense clumps useful as ground covers, as bank covers in moist shaded areas, and in hanging baskets. Fronds and parts of fronds are incorporated into leis and arrangements.

Calathea majestica var. *roseo-lineata*
ROSE-LINED CALATHEA

A wide-ranging species, this handsome plant is native to the northern rim of Amazonia from Brazil to Peru. Reaching 2 feet in height and forming dense clumps, it can be used as a ground or bank cover or as a foliage color accent.

Calathea makoyana
PEACOCK PLANT, CATHEDRAL WINDOWS

From Brazil, this strongly patterned species attains a height of 20 inches. It can be used as a single specimen, as a border, or as a ground or bank cover.

Calathea orbifolia
ROUND-LEAVED CALATHEA

A departure from the frequently seen long, narrow-leaved species, this Brazilian native displays its large, rounded foliage on an 18-inch plant. Use it as a bold accent or for a border or ground cover.

Calathea pseudoveitchiana
FISHTAIL CALATHEA

Displaying a startling pattern, this Brazilian species grows to 3 feet high. It is useful as an accent, border, or ground cover.

Calathea roseo-picta
ROSE-STREAKED CALATHEA

From Brazil, this species makes a dense ground cover only 8 inches high. It is also useful to form a border and makes a striking potted specimen.

Calathea roseo-picta 'Asian Beauty'
CALATHEA ASIAN BEAUTY

Originally a Brazilian species, this colorful sport or break occurred during tissue culture propagation. A low grower up to 8 inches high like its parent, it makes a unique ground cover and a real conversation piece as a potted specimen.

Calathea variegata
BANDED CALATHEA

Native to the northern and western rim of Amazonia, this richly patterned species reaches a height of 3 feet. Use it as a deep ground or bank cover or foliage accent.

Calathea zebrina 'Humilior'
CALATHEA HUMILIOR

Growing to 3 feet high, this Brazilian species can be used as a low shrub or tall ground cover. It forms dense clumps.

Ctenanthe burle-marxii
CTENANTHE BURLE-MARX

A spreading species from Brazil, this useful plant grows to 15 inches high. It is an excellent choice for a dense ground or bank cover in deep shade.

Ctenanthe oppenheimiana 'Tricolor'
NEVER-NEVER PLANT

A highly popular house plant, this Brazilian species displays richly patterned variegation. The underside of the leaf is light red purple, while the white variegation is seen above. Reaching 3 feet in height, it is useful as a low shrub, in a border, or as a deep ground cover.

Ctenanthe pilosa 'Golden Mosaic'
GOLDEN MOSAIC

One of the larger marantas, this showy cultivar reaches 5 feet in height and can be used as a low hedge, color accent, or as a deep ground cover under larger plants.

Maranta leuconeura var. *kerchoviana*
RABBIT TRACKS

Growing to 8 inches high or sometimes prostrate, this species from Brazil is an excellent ground or bank cover. It makes an interesting small potted accent.

Chapter 10
Bamboos, Cycads, and Palms

Bamboos

Bamboo is the collective term for some 1,500 species in the eighty genera representing this useful and beautiful group of plants. They are found worldwide, in both the New and Old Worlds, mostly in tropical and subtropical areas, although many are native to the eastern Asian temperate zone. Of widely varying sizes and habit, bamboo ranges from spreading ground covers only a few inches tall to giant construction types over 100 feet in height. There are two general types: runners and clumpers. The runners may become invasive and must be controlled by deep root barriers in the landscape. The clumpers form noninvasive clusters of densely packed culms or stems.

With one exception, we recommend only the clumpers for landscape use. All make excellent container specimens and can be grown indoors in well-lighted spaces. All are dramatic accents in the nightscape. Relative to other plant groups, bamboo is rapid growing. Rates of growth indicated below, therefore, indicate those compared with other species only within the bamboo group. Bamboo thrives in rich, moist, but well-drained soils. Most species prefer full sun. A member of the Grass Family (Poaceae), bamboo shares its usefulness with such other important grasses as wheat, corn, and rice—man's greatest staple foods.

Bambusa malingensis
MALING BAMBOO

Rapidly reaching 35 feet in height with 2-inch canes becoming yellow and green, this Chinese species with reputed strong salt air tolerance makes an excellent screen or windbreak in coastal areas with onshore winds. It can be used as a striking accent or background subject. The hard, straight wood is used in crafting implements and toys.

Bambusa multiplex 'Alphonse Karr'
CHINESE HEDGE BAMBOO

Tolerant of salt wind, this 25-foot high cultivar is excellent for a hedge, windbreak, or screen in coastal gardens. Its slender culms, forming very dense stands, are 1.5 inches in diameter. New culms are pink striped with green, later becoming all green. This Chinese cultivar may also be used as an accent specimen. It is a moderate grower.

Bambusa multiplex 'Silverstripe'
CHINESE SILVERSTRIPE HEDGE BAMBOO

With its green- and white-striped canes and foliage, this moderately fast growing Chinese cultivar reaches 25 feet in height and forms a dense clump used for screens and hedges. Its form makes it an excellent accent specimen for the small garden. It is wind, heat, and partially salt air tolerant.

Bambusa textilis

WEAVER'S BAMBOO

Popular in its native China, the fibrous canes of this moderate to rapid grower reach 40 feet in height, forming a compact, highly ornamental hedge, screen, or accent specimen. Its canes are used in basketry and similar craft work. A substance known as *tabisheer* is produced in cane internodes and prized in Chinese medicine.

Bambusa vulgaris vittata

GIANT GOLDEN BAMBOO, 'OHE NUI

Native to China, Giant Golden Bamboo has been a longtime favorite in Hawai'i's gardens. With canes 4 inches in diameter reaching 60 feet in height, it makes an excellent color accent, tall screen, and windbreak. It is a moderate grower that displays fair tolerance to salt air. Irritating hairs on the outside of leaf sheaths are to be avoided. The canes are used in arrangements.

Bambusa wamin

WAMIN BAMBOO, LUMPY NOODLE BAMBOO

A rather slow grower, this small Thai and Chinese bamboo reaches only 15 feet in height and displays fat, 3-inch canes heavily swollen at the internodes. It is similar to but more ornamental than the old-timer, Buddha's Belly Bamboo. It makes an interesting accent for a small garden space.

Chusquea coronalis

COSTA RICAN WEEPING BAMBOO

An exceptionally graceful, delicate species, this Central American bamboo grows to 20 feet high, making a fine specimen, hedge, or low screen for the small garden. It is a moderate grower. Costa Rican Weeping Bamboo does not thrive in soggy soils; the growing medium must be well drained.

Dendrocalamus brandesii

VELVET LEAF BAMBOO

A very rapid grower, this native of the area from Myanmar across to southern China produces heavy canes up to 8 inches in diameter that reach 60 feet in height. They are reputed to grow to 120 feet high under ideal conditions. The silvery canes are of timber quality, and the shoots are edible. A dramatic accent plant for the larger garden, park or campus, it is also used for tall screens and windbreaks.

Dendrocalamus membranaceus
PAI SAANG BAMBOO

A very rapid grower, Pai Saang Bamboo reaches 50 feet in height with occasionally striped 4-inch canes. This Burmese and Thai bamboo with its gracefully arching habit is a beautiful accent, canopy species, or tall windbreak and screen. It produces edible shoots. Young canes are used in basketry, older canes for construction.

Otatea acuminata aztecorum
MEXICAN WEEPING BAMBOO

With slender purplish canes only 1.5 inches in diameter but reaching 20 feet in height, this moderately fast growing Mexican species tolerates mild salt spray and short periods of drought. Its canes are solid and used in construction and the shoots are edible. Use this unusual plant as a screen, a background, or as a free-standing filigree-like accent in either sun or light shade.

Pleioblastus viride striata
DWARF STRIPED BAMBOO

Although a runner, this variegated Japanese species is a dwarf that grows slowly to only 1 foot in height with a 2-foot spread. It can thus be readily contained. It is an excellent ground cover plant in light shade and benefits from a winter mowing to 1 inch, which forces new growth with optimum variegation. It is also an excellent potted specimen.

Schizostachys glaucifolium
'OHE KAHIKO, HULA BAMBOO

Native to islands of the South Pacific, this Polynesian introduction produces slow growing 2-inch canes that reach 40 feet in height and form a relatively open clump. It makes a beautiful grove where the light foliage can shade a garden walk. *'Ohe Kahiko* canes are thin walled with long internodes, making them ideal for splitting. Traditionally, Hawaiians used them for musical instruments and *kapa* pattern stamps.

Thyrsostachys siamensis
MONASTERY BAMBOO

Grown in commercial Thai plantations for its excellent edible shoots and ornamental qualities, Monastery Bamboo canes reach 30 feet in height and are thick walled, straight, and strong, making them useful in construction. Clumps are compact and columnar, well suited for use in small gardens. It displays both wind and drought tolerance. It is native to Myanmar.

Bamboos, Cycads, and Palms

Cycads

Superficially, cycads resemble palms, many topping single trunks with palmlike foliage, others forming dense clusters. They are primitive members of the plant kingdom and were widespread during the Mesozoic era. Remnants of this ancient flora furnish excellent subjects for the tropical landscape. The term *cycads* encompasses three plant families: Cycadaceae, Stangeriaceae, and Zamiaceae. These in turn comprise eleven genera and probably well over a hundred species. Current interest in cycads is resulting in the search for, discovery, and naming of many new species.

Ceratozamia hildae
BAMBOO CYCAD

Zamiaceae (Coontie Family)

This Mexican cycad, with a trunk up to 5 feet high and a crown of stiff, pinnately compound leaves, makes a strong accent plant. An easy to grow evergreen plant, use it as an accent in a rockery or as tall ground cover for the partially shaded garden. It is a good potted specimen.

Ceratozamia mexicana
MEXICAN HORNCONE

Zamiaceae (Coontie Family)

A cycad from Mexico with a trunk up to 6 feet high, this graceful species is a useful accent or background plant. It also takes well to containerizing. Adapted to almost any well-drained soil, this cycad should be planted in partial shade.

Cycas circinalis
SAGO PALM, QUEEN SAGO (P) (T)

Cycadaceae (Cycad Family)

This slow growing Asian evergreen species develops both single and multiple trunks up to 20 feet high. It is best grown in full sun in a well-drained soil and displays good tolerance of wind, drought, and salt. It will tolerate partial shade. Easily grown, it can be used as a specimen or potted for the lanai. The foliage may be cut for arrangements.

Cycas revoluta
SAGO PALM, JAPANESE SAGO PALM, KING SAGO (P) (T)

Cycadaceae (Cycad Family)

This very slow growing evergreen cycad from southern Japan develops a stout, 10-foot high trunk with many offshoots and a dense crown of glossy leaves that may be cut for arrangements. Easy to grow in sun or part shade in a well-drained soil, it displays moderate tolerance to drought and good tolerance to wind and salt. An excellent potted plant, it is a strong, almost indestructible accent in the landscape.

Dioon spinulosum
SMALL-SPINED CYCAD
Zamiaceae (Coontie Family)

A tall Mexican cycad from wet forests, this species reaches a height of 30 feet or more. A slow grower, it may be used in the cycad garden as a canopy plant, specimen, as background, or as a tubbed accent for the lanai. It is moderately wind tolerant with some salt air tolerance.

Encephalartos lehmannii
KAROO CYCAD (S)
Zamiaceae (Coontie Family)

With its bright foliage heightened by full sun, the Karoo Cycad is a commanding accent plant in the general landscape and makes an excellent tubbed specimen. It is highly tolerant of heat and must have a well-drained soil. It is partially drought tolerant, salt tolerant, and completely wind tolerant. Slow growing, its trunk may attain a height of 6 feet. It is native to the East Cape of South Africa. Its color makes it a good subject for the nightscape.

Encephalartos villosus
WOOLY CYCAD
Zamiaceae (Coontie Family)

A small, trunkless cycad from Swaziland to the East Cape of South Africa, this tough plant performs well in moist, partially shaded areas. Use it as an understory plant. It is also suitable for use on the shaded lanai or terrace in a container.

Macrozamia macdonnellii
MACDONNELL RANGE CYCAD
Zamiaceae (Coontie Family)

Planted in full sun, the whorl of bright leaves of this slow growing species is a strong focal point in both the landscape and nightscape. It can also be used as a tubbed accent plant on a sunny lanai or deck. Many leaves up to 4 feet long crown a trunk that may reach a height of 10 feet. Native to hot, arid areas of central Australia, this cycad is heat and drought tolerant and has some tolerance of salt air. It requires a well-drained soil and moderate watering.

Stangeria eriopus
HOTTENTOT'S HEAD
Stangeriaceae (Stangeria Family)

This unique cycad from eastern South Africa is in a family by itself, with but one genus and one species. It is fernlike with leaves up to 6 feet long. Slow growing, Stangeria requires light shade and a moist, rich, well-drained soil. It has no heat, drought, or salt tolerance. Use it as an understory foliage accent or tubbed on a shady lanai.

Zamia furfuracea
CARDBOARD PALM (P)
Zamiaceae (Coontie Family)

This native of Mexico's southeastern coast does best in full sun but will tolerate heavy shade. It is heat, wind, and partially drought tolerant. It grows well in any well-drained garden soil and will take happily to sandy soils near the ocean, where it exhibits tolerance to salt air. Its mounded growth will reach 8 feet in height. It is an excellent potted plant, specimen, or ground and bank cover and finds use in the interiorscape.

Zamia integrifolia
COONTIE, SEMINOLE BREAD, SAGO CYCAS (P)
Zamiaceae (Coontie Family)

Found in dune vegetation and adjacent partially shaded habitats from Florida into the West Indies, Coontie is best used as a shrubby ground cover as well as a tough potted specimen for the lanai or terrace. It thrives in sandy limestone soils but grows readily in any good garden soil. Fine-textured leaves crown a trunk 2 feet high.

Palms

Universally associated with tropical gardens, palms rank among the most highly prized of all plants for their unique form and grace and their usefulness in the landscape. Palms vary greatly, from trees 200 feet high to low shrubs and even vines. Mostly tropical, there are about 185 genera and possibly as many as 2,500 species distributed worldwide. Many are of high economic importance, furnishing oils, beverages, and construction materials. Palm taxonomy is under constant scrutiny and revision. All palms are placed in the Arecaceae—the Palm Family.

Acoelorraphe wrightii
PAUROTIS PALM, EVERGLADES PALM (T)

An attractive, many-trunked, slender palm from the Florida Everglades, the Paurotis Palm grows in full sun to a height of 40 feet. It requires damp to wet soils and regular applications of fertilizer for best results. Use this palm as a striking garden accent, hedge, or thick screen. Leaves are utilized in arrangements.

Archontophoenix alexandrae
ALEXANDRA PALM, KING PALM, KING ALEXANDER PALM

A stately, solitary Australian palm up to 90 feet high, this palm requires moisture but will grow in almost any soil. It has poor salt tolerance. Clusters of white flowers are followed by masses of bright red fruit. Use it for street or avenue plantings or in groupings, as a background accent or, when small, as a potted specimen in either a shady or sunny situation.

Bismarckia nobilis
BISMARCK PALM

From drier parts of Madagascar, this species is ideal for Hawai'i's leeward gardens. It is very slow growing, eventually reaching 200 feet in height. The large silvery leaves provide a major accent in the home landscape or in parks and large developments. Its bright leaves can be used in cut arrangements and are spectacular in the nightscape. Grow in full sun in a hot situation and water moderately.

Brahea armata
BLUE HESPER PALM (T)

The leaves of this Mexican species attain their best color in full sun in a warm situation with low rainfall. Only moderate watering is required. A fairly slow grower, it will eventually reach 40 feet in height. Use it as a color accent or a potted specimen for the sunny lanai. Foliage may be used for arrangements and provides an accent in the nightscape.

Carpentaria acuminata
CARPENTARIA PALM

This solitary palm from Australia grows moderately fast to a height of 90 feet. Its clusters of small red fruit may be cut for arrangements. It requires full sun in a moist soil and is not tolerant of salts. It makes a beautiful grove or avenue tree and takes readily to containerizing.

Caryota mitis
FISHTAIL PALM, CLUSTERING FISHTAIL PALM (S)

A densely clumping palm from Southeast Asia used as a landscape accent specimen, a screen, or a tall hedge, this species attains a height of 40 feet. It performs best in full sun or light shade in moist, rich soils. It makes an excellent tubbed specimen and can be used in the interiorscape. It has no drought or salt tolerance.

Caryota urens
WINE PALM, TODDY PALM, JAGGERY PALM (S)

This solitary, rapidly growing palm is a temporary landscape specimen, lasting about ten years prior to declining. It has been successfully used to provide quick, luxuriant growth while slower growing specimens are being established. Attaining a height of 40 feet in a good moist soil, it can be used as an accent, a screen, or windbreak. It prefers full sun.

Bamboos, Cycads, and Palms

Chamaedorea brachypoda
SHORT-STALKED CHAMAEDOREA

Native to Guatemala and Honduras, this is a densely clumping species attaining a height of 6 feet and forming clumps several times that across. It requires shade and moisture, where it finds good use as a bushy ground cover, bank cover, or general understory subject. It flourishes as a potted specimen and can be used indoors in a bright spot or outside on the lanai.

Chamaedorea cataractarum
CASCADE PALM, CAT PALM

Mexico provides the native habitat of this densely clumping species. It may attain a height of 6 feet, spreading to 9 feet. It is useful as a hedge, screen, border, or general understory in a partially shaded, moist situation. It makes a fine tubbed specimen for indoor use or for the lanai. It does best protected from drying winds and has no drought or salt tolerance.

Chamaedorea metallica
METALLICA, MINIATURE FISHTAIL PALM

This single-trunked Mexican palm is best as an understory shrub in the moist, shaded garden. Growing to 9 feet high, it provides an accent either alone or in groups. It makes an excellent potted specimen for sheltered, shaded areas. It has no drought or salt tolerance and does best when protected from harsh winds.

Chamaedorea microspadix
HARDY BAMBOO PALM

A clustering palm from Mexico that grows to 10 feet high, this species is used in the moist, shaded landscape away from wind as an accent, hedge, or screen, or it can be grown as a potted specimen. It grows well indoors if given strong light. The clusters of bright orange red fruits add a color accent and can be used in arrangements.

Chamaedorea seifrizii
BAMBOO PALM, REED PALM

This is a dense, moderately fast growing, clumping species from Mexico. It reaches 8 feet in height and requires high moisture, shade, and protection from wind. Use it in the garden understory as a screen or hedge or as a potted specimen either indoors or on the lanai. The black fruits carried on bright orange stems are a color accent and may be used in arrangements.

Chambeyronia macrocarpa
BLUSHING PALM

A somewhat rare, slow growing, solitary palm from New Caledonia, this species attains a height of 60 feet under conditions of good soil, frequent watering, and regular feeding. It prefers full sun but will tolerate light shade. New leaves are bright red, gradually turning green. It makes a good canopy plant or an accent where its new leaves can be seen. It has no tolerance of salt.

Coccothrinax argentea
SILVER PALM

The Silver Palm is a slow growing, solitary palm from Florida and the Bahamas with a slender, light gray trunk. This is a very tough and durable fan palm up to 20 feet high, densely silver on the underside of its green leaves. It is a useful small specimen for the garden, good for seaside plantings, and suitable for container use indoors or on the lanai. Flowers, foliage, and fruit are used in arrangements. It is a bright accent in the nightscape.

Cocos nucifera
COCONUT PALM, NIU

The Coconut Palm is the universal symbol of the tropics—and the world's most economically important palm. It is a moderately fast growing plant that reaches a height of 100 feet. The large, edible coconut, whose husk may be green, yellow, or orange, as well as the flowers, flower sheath, and leaves, are used in arrangements. It has good salt tolerance at the root zone, but the leaves are subject to salt spray damage and will burn if not rain-washed. They are excellent for large gardens, parks, or avenues. They need ample moisture.

   A smaller form known locally as the Samoan Coconut displays shorter, broader leaves and bears larger, rounded, bright green fruit. It may reach 40 feet in height. The Samoan form is recommended for the small- to medium-sized garden.

Copernicia prunifera
CARNAUBA WAX PALM, WAX PALM (T)

With its bright, silvery, palmate leaves, the Carnauba Wax Palm provides an almost luminous accent in the garden, both in the daytime and the nightscape. Native to Brazil, its solitary trunk may reach a height of 40 feet. Grow it in full sun with moderate watering and regular feeding. Leaves are used in arrangements.

Bamboos, Cycads, and Palms

Cyrtostachys renda
SEALING WAX PALM, RED SEALING WAX PALM, LIPSTICK PALM

This clumping palm up to 30 feet high comes from Malaysia and Borneo. It displays a neon red crown shaft, which renders it an excellent accent specimen. Use it as a hedge or screen or as a potted accent. It requires moist soils and does well along a stream or pond; it will not tolerate drought, salt, or drying winds. Its red topped stems may be used in arrangements.

Dictyosperma album
PRINCESS PALM, RED PALM, HURRICANE PALM

A slender, single-trunked palm native to the Mascarene Islands including Mauritius, this species will grow to 40 feet high. It tolerates light shade but performs best in full sun in good soil with ample water. It will grow well in a lanai container if regularly watered. It prefers protection from strong wind. The Princess Palm is excellent planted in grove-like groupings.

Dypsis decaryi
TRIANGLE PALM, MADAGASCAN THREE-SIDED PALM

A Madagascan species, this palm is slow growing to 35 feet high. It is adapted to most high light locations and has moderate drought tolerance. It is not tolerant of salt but withstands heat and wind. Its unique form makes it an excellent specimen for the garden and for container use. It is an attractive nightscape subject.

Dypsis lutescens
GOLDEN-FRUITED PALM, ARECA PALM

As an attractive specimen, a screen, hedge, windbreak, boundary plant, or as a potted specimen for the lanai or indoors, this clumping palm from Madagascar has no peer. Growing to 30 feet high, it requires good soil and ample watering. It produces attractive clusters of golden yellow fruit used in arrangements.

Howea forsteriana
SENTRY PALM, KENTIA PALM

Australia's Lord Howe Island is home to this slow growing, single-trunked palm widely used as a house plant. It will tolerate low light and long-term potting. Planted in the garden, it will grow to 40 feet high. It prefers good, well-drained soils with regular watering and feeding and provides a useful accent or grouping palm.

Hyophorbe lagenicaulis
BOTTLE PALM

Popular for its unusual trunk, this small, slow growing palm may reach a height of 10 feet. It likes full sun and moderate watering. It will do very well as a potted or tubbed specimen. Because of its somewhat startling shape, it is not usually used in groupings but rather as an accent. The Bottle Palm is native to the Mascarene Islands east of Madagascar.

Latania loddigesii
BLUE LATAN PALM

This slow growing species from the Mascarene Islands may attain a height of 40 feet. Its foliage provides a colorful accent in the landscape and nightscape. It responds to full sun, moderate watering, and a well-drained soil. Its masses of greenish brown fruit are used in arrangements, as are the large leaves.

Licuala grandis
LICUALA PALM (T)

This small species from New Britain is best in light shade and a good, well-drained soil with ample moisture. The large leaves, topping a slender trunk up to 10 feet high, require shelter from wind. It makes an excellent tubbed specimen or accent planted either as a single specimen or in groves. Sprays of its orange red fruit and large leaves may be used in arrangements.

Licuala spinosa
SPINY LICUALA PALM (T)

A useful, densely clumping palm from Java and the Moluccas, this species may reach 10 feet in height. It does best in wet soils such as pond edges and streams, even close to brackish water, in either full sun or part shade. Protection from strong winds prevents leaf damage. Use as a hedge, screen, or barrier planting. Sprays of its orange fruit and large leaves are used in arrangements.

Livistona chinensis
CHINESE FAN PALM, FOUNTAIN PALM (T)

This slow growing solitary palm up to 30 feet high is native to China. Best grown in full sun in most soils, it has fair salt and moderate drought tolerance. Use it as a specimen, a background border, or clumped as an accent where it will receive water. It is wind resistant. Masses of its attractive blue green fruit may be used in arrangements.

Normanbya normanbyi
Black Palm

A single-trunked, slender palm reaching 60 feet in height, the Black Palm is native to Australia. It prefers good, moist soils and is a moderate grower. It is useful as a specimen, in groups for forming a canopy for understory plants, or as a tubbed specimen.

Phoenix roebelenii
Dwarf Date Palm, Pygmy Date Palm (T)

This solitary palm from Laos is slow growing to 10 feet high. Easily grown, it is a good accent used individually or massed in a grove, in planter boxes, or as an interior plant. It thrives in full sun or part shade and needs good, well-drained soil and ample water. It responds to regular fertilizing.

Pinanga kuhlii
Ivory Cane Palm

This beautiful clustering palm is native to wet forest areas of Sumatra and Java, where it prefers the shaded understory. It will take to pot culture and can be used in the landscape as an accent specimen or grouped for screening. It may attain a height of 30 feet and is a fairly rapid grower, given ample water and feeding.

Pritchardia hillebrandii
Loulu Lelo

This native Hawaiian species is a slow growing, solitary palm up to 20 feet high. It does best in full sun, which brings out the bluish silver color of its foliage. It prefers a well-drained, moderately moist soil, and makes a striking color accent in the landscape and nightscape. It will tolerate container culture. The leaves, the yellow-flowered inflorescences, and the clusters of black fruits are used in arrangements. It is one of over two dozen endemic Pritchardia species.

Pritchardia pacifica
Fiji Fan Palm

This Fijian palm is a moderately fast growing solitary species up to 30 feet high. Growing in full sun with even moisture and protection from drying winds, it is useful as a framing tree, an accent, or planted in a grove and is suitable as a potted plant when young. It has moderate salt tolerance.

Pritchardia thurstonii
THURSTON FAN PALM

This slow growing solitary palm from Fiji makes an attractive accent plant. It is best in full sun with good, well-drained soil. It has good salt tolerance. Reaching 30 feet in height, it makes a beautiful specimen plant and is useful planted in groves. It is suitable as a potted plant in the interior or on the lanai.

Ptychosperma macarthurii
MACARTHUR PALM (S)

A clumping palm growing to 25 feet high, this palm prefers light shade, a rich, moist, well-drained soil, and protection from harsh winds. It has no tolerance to drying or salt exposure. Use it as a screen or hedge or a large understory subject. It does well in containers, tolerating considerable shade as an indoor plant. New Guinea is its native land.

Rhapis excelsa
LADY PALM, BAMBOO PALM

This slow growing clumping palm, 10 feet high, originates in southern China. It is excellent for shaded locations but will tolerate sun in most well-drained soils. It has moderate drought and salt tolerance. It makes an excellent specimen and is useful planted as a hedge or screen. It is an excellent potted plant for the deck, lanai, or interior. Leaves may be used in arrangements.

Roystonea oleracea
CABBAGE PALM, CARIBBEE ROYAL PALM

Best grown in full sun in a moist, well-drained soil, this stately palm has good wind and salt resistance. It is used as a street or avenue tree, for framing large buildings, or for major palm groupings. Rather fast growing, the trunk may attain a height of over 100 feet. It is native to the Caribbean and South America. The flower sheath is used in arrangements.

Roystonea regia
ROYAL PALM, CUBAN ROYAL PALM

This majestic, fast growing palm from Cuba, grown in full sun in a well-drained soil with ample water, can reach a height of 70 feet. It is wind resistant and has moderate drought and salt tolerance. It is useful as a street or avenue tree, for framing large buildings, or for palm groupings. The flower sheath is used in arrangements.

Bamboos, Cycads, and Palms

Syagrus romanzoffianum
Queen Palm, Monkey Nut Palm

A native of southern Brazil, this graceful palm is a moderate grower in rich, moist, well-drained soil and may reach a height of 50 feet. It may be used as a strong vertical accent in the small garden; it has also been used as a street tree or a formal avenue plant.

Thrinax parviflora
Jamaican Thatch Palm

This slow growing solitary palm from Jamaica has a slender gray trunk with fiber in the upper portion. Growing to 20 feet high, it has a light green, fan-shaped leaf. It may be used in sun or light shade in most soils and has a high tolerance to salts and alkaline soils. This easily grown palm is a suitable accent for the small garden or as a container plant.

Veitchia merrillii
Manila Palm, Christmas Palm, Merrill Palm

From the Philippines, this is a moderately fast growing solitary palm up to 20 feet high. It is useful in the small garden as a specimen or as a container plant for the lanai or interior. It thrives in sun or moderate shade in most soils and has moderate drought and salt tolerance. The clusters of bright red fruit, appearing during the holiday season, may be used in arrangements.

Washingtonia robusta
Mexican Fan Palm, Washington Palm, Washingtonia Palm (T)

This large, fast growing Mexican species attains a height of 100 feet. It is suitable as a street or avenue tree or as a specimen plant for larger properties; sometimes it is used in cluster plantings. Best grown in full sun in most well-drained soils, it has moderate tolerance to salt.

Wodyetia bifurcata
Foxtail Palm

Northern Queensland, Australia, is the home of this rapidly growing species. Of easy growth in a well-drained, moderately moist soil, it reaches a height of 30 feet. It displays some drought tolerance. In the home landscape, use it in a grove planting to provide a rapid canopy for shade lovers. It makes a graceful single specimen in the garden or in a container.

Appendix A: Landscaping with Native Hawaiian Species

The successful use of native species in the landscape has been demonstrated only in recent years, although gardeners have long planted the Hawaiian tree ferns, Bird's-Nest Fern, *Palai*, and *Naupaka*. There are many other species available now from local plant sales. We take a strong conservation stance: Do not remove native plants from the wild. Those of landscape value will appear in botanical garden sales propagated from material carefully and judiciously collected from the wild, grown in the botanical garden, propagated from that cultivated material, and finally released for public use. No damage to the native flora results.

A number of highly ornamental endangered species are sold under special State permits. In the following listing, these are indicated by (E) following the scientific name.

Although there is much yet to learn about growing native Hawaiian species, several rules of thumb are in order. Native species on the whole demand excellent drainage and soil aeration. Species requiring moisture must be kept moist, but they succumb rapidly to stagnant, soggy soils. Most require only minimal mulching; some none at all. In general, coastal species can do without mulch, while mountain species welcome a 2-inch layer. If a species requires sun, give it sun and if not, avoid it. Fertilize conservatively. Many native plants decline rapidly if overdosed with well-intended nutrients. Many are highly sensitive to horticultural chemicals, the preemergents, herbicides, insecticides, and fungicides. Act conservatively. Those species marked with an asterisk (*) are not included in the text but may be found in the list of suggested readings.

Ground Covers:
Bacopa monnieri
Heliotropium anomalum var. *argenteum*
Ipomoea pes-caprae subsp. *brasiliensis*
Jaquemontia ovalifolia subsp. *sandwicensis*
Lipochaeta integrifolia
Peperomia leptostachya *
Plectranthus parviflorus *
Plumbago zeylanica
Portulaca molokiniensis * (E)
Scaevola coriacea * (E)
Sida fallax
Vitex rotundifolia

Small Shrubs:
Achyranthes splendens
Argemone sandwicensis *
Artemisia mauiensis
Astelia menziesiana *
Bidens spp. *
Brighamia insignis *(E)
Brighamia rockii * (E)
Chamaesyce celastrides var. *celastroides* *
Sesbania tomentosa * (E)
Wikstroemia pulcherrima *
Wikstroemia uva-ursi

Medium Shrubs:
Bidens micrantha subsp. *ctenophylla* * (E)
Gossypium tomentosum

Hibiscus brackenridgei * (E)
Hibiscus clayi * (E)
Hibiscus kokio *
Nototrichium sandwicensis
Osteomeles anthyllidifolia
Scaevola gaudichaudii *
Scaevola taccada
Sida fallax

Large Shrubs:
Dodonaea viscosa
Dracaena aurea *
Hibiscus arnottianus subsp. *Immaculatus* * (E)
Sophora chrysophylla *

Small Trees:
Gardenia brighamii * (E)
Hibiscus arnottianus
Hibiscus arnottianus subsp. *punaluuensis*
Hibiscus waimeae *
Myoporum sandwicense
Pittosporum hosmeri
Psydrax odorata

Medium Trees:
Acacia koa
Caesalpinia kavaiensis * (E)

Erythrina sandwicensis
Metrosideros polymorpha
Reynoldsia sandwicensis *
Sapindus saponaria
Sapindus oahuensis *
Tetraplasandra hawaiiensis *

Vines:
Alyxia oliviformis
Canavalia galeata *

Ferns:
Adiantum capillus-veneris
Asplenium nidus
Cibotium chamissoi *
Cibotium glaucum
Marsilea villosa * (E)
Microlepia strigosa
Nephrolepis corditolia
Nephrolepis exaltata
Odontosoria chinensis
Psilotum nudum
Sadleria cyatheoides *

Palms:
Pritchardia affinis * (E)
Pritchardia arecina *
Pritchardia hillebrandii
Pritchardia martii *

Heritage Species

There is a special group of landscape-worthy plants commonly seen in Hawaiian gardens. Although considered Hawaiian, they were introduced during the period of Polynesian migration. These are known as *Heritage Species*, and they include a number of species of landscape interest.

Small Shrubs:
Colocasia esculenta and cultivars *

Large Shrubs:
Cordyline fruticosa
Piper methysticum *
Saccharum officinarum and cultivars *

Small Trees:
Morinda citrifolia

Medium Trees:
Aleurites moluccana
Cordia subcordata
Hibiscus tiliaceus
Pandanus tectorius

Syzygium malaccense
Thespesia populnea

Large Trees:
Artocarpus altilis
Calophyllum inophyllum

Palms:
Cocos nucifera

Appendix B: Plants for the Xeriscape

Derived from a Greek word meaning "dry," *Xeriscape* has become a popular term for wise use of water in the landscape. It involves using plants which, after a period of establishment, will continue to thrive with little or no added irrigation. There are many other factors enhancing water conservation in the garden, but the scope of this publication addresses only plant species adapted to the dry tropics, which in Hawai'i covers all of the leeward coast areas. In the following list, species marked with a plus (+) are highly drought tolerant once established. Those species marked with an asterisk (*) are not included in the text but may be found in the list of suggested readings. The list is not complete, and the gardener—through trial and error—may add to the list as dictated by personal experience.

Ground Covers:
Aloe vera +
Aloe spp. + *
Asparagus densiflorus 'Myers'
Asparagus densiflorus 'Sprengeri'
Asystasia gangetica +
Catharanthus hybrids *
Heliotropium anomalum var. argenteum +
Jaquemontia ovalifolium subsp. sandwicensis +
Lantana montevidensis +
Lantana montevidensis cultivars + *
Lipochaeta integrifolia +
Plumbago zeylanica +
Rosmarinus officinalis +
Sida fallax (creeping form) +
Tradescantia pallida 'Purpurea'
Tradescantia spathacea
Tradescantia spathacea 'Dwarf'
Verbena x hybrida
Vitex rotundifolia +
Zamia furfuracea

Small Shrubs:
Achyranthes splendens +
Aloe spp. + *
Argemone sandwicensis + *
Carissa macrocarpa cultivars + *
Chamaecyce celastroides var. celastroides *
Crassula ovata +
Crassula argentea cultivars + *
Euphorbia milii +
Euphorbia spp.+ *
Plectranthus amboinicus + *
Sansevieria trifasciata +
Sansevieria trifasciata cultivars +
Sansevieria spp. + *
Senecio cineraria
Turnera ulmifolia
Wikstroemia uva-ursi

Medium Shrubs:
Adenium multiflorum + *
Adenium obesum +
Adenium swazicum + *
Agave attenuata +
Agave spp. + *
Allamanda schotti
Anisacanthus thurberi
Cortederia selloana + *
Dasylirion longissimum +
Dasylirion wheeleri +
Furcraea spp. + *
Gardenia volkensii *

Gossypium tomentosum +
Hibiscus rockii (H. calophyllus) + *
Hibiscus scottii + *
Jasminum sambac
Kalanchoe beharensis +
Kalanchoe spp. + *
Lantana camara +
Lantana camara cultivars + *
Leucophyllum frutescens
Nolina recurvata +
Nototrichium sandwicense +
Osteomeles anthyllidifolia
Plumbago auriculata
Plumbago auriculata var. alba
Portulacaria afra + *
Portulacaria afra cultivars + *
Punica granatum +
Rosmarinus officinalis +
Russellia equisetifolia
Scaevola taccada +
Sida fallax +
Synadenium compacta var. rubra + *
Synadenium granti + *

Large Shrubs:
Calotropis gigantea +
Calotropis procera + *
Carissa macrocarpa +
Dodonaea viscosa +
Furcraea macdougalii + *
Lysiloma thornberi +
Nerium oleander +

Small Trees:
Bauhinia binata
Bauhinia tomentosa
Caesalpinia pulcherrima +
Caesalpinia pulcherrima 'Compton' + *
Caesalpinia pulcherrima forma flava +
Clusia rosea +
Coccoloba uvifera
Crescentia alata *
Crescentia cujete
Cupaniopsis anacardioides *
Cussonia paniculata *
Cussonia spicata *
Dracaena draco *
Ficus carica + *
Ipomoea arborescens + *
Morinda citrifolia +
Myoporum sandwicensis +
Parkinsonia aculeata + *
Psydrax odorata +
Schinus terebinthifolius +
Senna surratensis +

Thevetia peruviana +
Yucca aloifolia + *
Yucca elephantipes +
Yucca elephantipes 'Variegata' + *

Medium Trees:
Azadarachta indica +
Bursera simarouba *
Caesalpinia ferrea *
Cupressus sempervirens
Delonix regia
Erythrina abyssinica *
Erythrina sandwicensis +
Erythrina vespertilio *
Ficus palmeri + *
Ficus petiolaris + *
Harpephyllum caffra *
Kigelia pinnata +
Melaleuca quinquenervia
Melia azedarach
Olea europaea subsp. europaea +
Pimenta dioica
Pithecellobium dulce 'Variegata'
Prosopis pallida +
Reynoldsia sandwicensis + *
Schinus molle +

Large Trees:
Casuarina equisetifolia
Enterolobium cyclocarpum
Ficus macrophylla
Ficus microcarpa +

Vines:
Antigonon leptopus +
Bauhinia galpinii
Canavalia galeata + *
Cissus rotundifolia + *
Cissus quadrangularis + *
Cryptostegia madagascariensis +
Jasminum officinalis *
Macfadyena unguis-cati
Marsdenia floribunda
Pseudogynoxys chenopodioides

Bromeliads:
Aechmea blanchettiana
Ananas comosus var. variegatus
Tillandsia balbisiana *
Tillandsia capitata 'Peach' *
Tillandsia concolor *
Tillandsia plumosa *
Tillandsia rodriguesiana *
Tillandsia xerigraphica
Vriesia imperialis

Appendix C: Plants for the Beach Garden

There are many species adapted to living under coastal conditions. Strong and salt-laden onshore winds—and actual salt spray—are basic factors preventing the use of most common landscape species in a beach garden. Such factors are intensified in a site receiving low annual rainfall. Using those plants with the greatest resistance to salt is critical.

The following list suggests a range of species for the beach garden. Those marked with a plus sign (+) are especially salt tolerant. Others grow better in the lee of a sheltering house or of species having greater salt tolerance. Those marked with an asterisk (*) are not included in the text but may be found in the list of suggested readings.

The environment of each garden varies, and the following list may be expanded through the gardener's experience.

Ground Covers:
Agapanthus praecox subsp. *orientalis*
Aloe spp. + *
Aloe vera +
Asparagus densiflorus 'Myers'
Asparagus densiflorus 'Sprengeri'
Bacopa monnieri +
Carpobrotus edulis + *
Gaillardia pulchella var. *picta* + *
Gaillardia hybrids + *
Gnphalium sandwicense + *
Heliotropium anomalum var. *argenteum* +
Ipomoea pes-caprae subsp. *brasiliensis* +
Jaquemontia ovalifolia subsp. *sandwicensis* +
Juniperus procumbens +
Lantana montevidensis +
Lantana montevidensis cultivars + *
Lipochaeta integrifolia +
Portulaca grandiflora +
Portulaca lutea + *
Portulaca molokiniensis + *
Sida fallax (creeping form) +
Tradescantia pallida 'Purpurea'
Tradescantia spathacea
Tradescantia spathacea 'Dwarf'
Turnera ulmifolia
Verbena x *hybrida*
Vitex rotundifolia +
Wedelia triloba +

Small Shrubs:
Achyranthes splendens
Agapanthus praecox subsp. *orientalis*
Carissa macrocarpa cultivars + *
Chamaecyse celastroides var. *celastroides* + *
Clerodendrum inerme +
Colobrina asiatica + *
Crassula ovata +
Euphorbia lactea + *
Euphorbia millii
Ficus 'Green Island' + *
Hemerocallis spp. *
Hibiscus hamabo *
Pittosporum tobira 'Wheeler's Dwarf' +
Portulacaria afra + *
Sansevieria trifasciata +
Sansevieria trifasciata 'Laurentii' +
Sansevieria trifasciata spp. + *
Wikstroemia uva-ursi +

Medium Shrubs:
Adenium multiflorum + *
Adenium obesum +
Adenium swazicum + *
Agave americana + *
Agave attenuata +
Agave sisalina + *
Capparis sandwichiana + *
Crinum asiaticum +
Crinum augustum + *
Lantana camara +
Juniperus chinensis
Kalanchoe beharensis
Lantana camara cultivars + *
Nototrichium sandwicense
Osteomeles anthyllidifolia +
Pittosporum tobira 'Variegata'
Rhaphiolepis indica
Rhaphiolepis var. *integerrima*
Russelia equisetifolia
Scaevola taccada +
Sophora tomentosa + *
Strelitzia reginae
Tecoma capensis

Large Shrubs:
Calotropis gigantea +
Calotropis procera + *
Carissa macrocarpa +
Dodonaea viscosa
Ficus microcarpa var. *crassifolia* +
Gardenia taitensis +
Ligustrum japonicum
Ligustrum japonicum 'Rotundifolium'
Ligustrum ovalifolium *
Nerium oleander
Pittosporum tobira +

Small Trees:
Bucida molineti
Caesalpinia pulcherrima
Clusea rosea +
Coccoloba uvifera +
Cordia sebestena
Dracaena draco *
Guaiacum officinale
Heritiera littoralis
Hibiscus tiliaceus
Hibiscus titiaceus var. *sterile*
Morinda citrifolia +
Myoporum sandwicense +
Pandanus baptisii + *
Pandanus dubius + *
Pandanus tectorius +
Parkinsonia aculeata + *
Piscidia piscipula *
Plumeria obtusa
Plumeria rubra and cultivars
Psidium cattleianum
Psidium cattleianum forma *lucidum*
Psydrax odoratum
Scaevola taccada +
Scaevola taccada (form from Nukulau Island, Fiji) + *
Schinus terebinthifolius
Senna surratensis
Tournefortia argentea +
Yucca aloifolia + *
Yucca elephantipes
Yucca elephantipes 'Variegata' *

Medium Trees:
Bucida buceras
Cerbera manghas *
Conocarpus erectus +
Cordia subcordata
Guettarda speciosa *
Lagunaria pattersonii *
Maniklara zapota *
Melaleuca quinquenervia
Melia azedarach
Mimusops caffra * +
Noronhia emarginata +
Ochrosia elliptica *
Ochrosia oppositifolia *
Prosopis pallida +
Simarouba glauca *
Swietenia mahogani
Tamarix aphylla + *
Terminalia catappa +
Thespesia populnea
Thevetia peruviana

Large Trees:
Barringtonia asiatica *
Calophyllum inophyllum
Casuarina equisetifolia +
Ficus macrophylla
Ficus microcarpa
Hernandia nymphaeifolia *
Melaleuca quinquenervia

Vines:
Canavalia galeata + *
Canavalia sericea + *
Cissus quadrangularis + *
Cissus rotundifolia + *
Cryptostegia madagascariensis
Ipomoea carica + *

Marsdenia floribunda
Pentalinon lutea
Vigna lutea [+] [*]

Bromeliads:
Aechmea blanchetiana 'Lemon'[*]
Aechmea blanchetiana 'Orange'[*]

Bamboos:
Bambusa malingensis
Bambusa multiplex 'Alphonse Karr'
Bambusa multiplex 'Silverstripe'
Bambusa vulgaris vitatta
Otatea acuminata aztecorum

Cycads:
Cycas circinnalis
Cycas revoluta
Encephalartos lehmannii
Zamia furfuracea
Zamia integrifolia

Palms:
Acoelorraphe wrightii
Coccothrinax argentata
Cocos nucifera
Hyophorbe lagenicaulis
Hyophorbe verschaffeltii [*]
Pritchardia hillebrandii
Pritchardia pacifica
Pritchardia thurstonii
Thrinax parviflora
Washingtonia robusta

Appendix D: Plants for Hedges, Screens, and Windbreaks

Protection from wind and the desire for privacy are common factors in designing a landscape. These needs are satisfied by a wide range of plants, from large trees to medium-high shrubs. Windbreaks must produce tough foliage resistant to the buffeting of wind and a habit of growth that provides foliage mass from canopy to nearly ground level. Hedges and other privacy plantings need not necessarily be wind tolerant, but they must have the top-to-bottom foliage character. They are used to screen out undesirable factors—a view of the street, a dog run, the neighbors, the compost pile, refuse containers—or simply to define or enclose special garden-use areas. The following list provides a number of choices. Many of the shrubs may be clipped into formal hedges if required for design purposes. Those species marked with an asterisk (*) are not covered in the text but may be found in the list of suggested readings.

Medium Shrubs:
Abelia x grandiflora
Allamanda blanchettii
Allamanda schottii
Ficus microcarpa var. crassifolia
Gardenia augusta
Graptophyllum pictum
Graptophyllum pictum 'Eldorado'
Ixora coccinea
Ixora casei 'Super King'
Ixora chinensis 'Nora Grant'
Leucophyllum frutescens
Mussaenda frondosa
Nototrichium sandwicense
Odontonema cuspidatum
Olea europaea subsp. europaea 'Little Ollie' *
Osteomeles anthyllidifolia
Pachystachys lutea
Pittosporum tobira 'Variegata'
Plumbago auriculata
Polyscias fruticosa
Pseuderanthemum carruthersi var. atropurpureum
Pseuderanthemum carruthersi var. reticulatum
Pseuderanthemum carruthersi var. variegatum
Rhaphiolepis indica
Rhaphiolepis umbellata var. integerrima
Sanchesia speciosa
Scaevola taccada
Tecoma capensis
Tecoma capensis 'Aurea'
Thunbergia erecta
Thunbergia erecta 'Alba'
Triphasia trifoliata⁺

Large Shrubs:
Breynia disticha 'Roseopicta'
Calotropis gigantea
Calotropis procera *
Carissa macrocarpa
Codaieum variegatum
Cordyline fruticosa
Cordyline fruticosa cultivars
Dodonaea viscosa
Duranta erecta
Eugenia uniflora
Gardenia taitensis
Gardenia volkensii *
Hibiscus rosa-sinensis
Hibiscus rosa-sinensis cultivars
Hibiscus schizopetalus
Hibiscus schizopetalus cultivars
Holmskioldia sanguinea
Holmskioldia sanguinea 'Citrina'
Leea guineensis
Ligustrum japonicum
Ligustrum japonicum 'Rotundifolium'
Malvaviscus penduliflorus
Murraya paniculata
Mussaenda x 'Doña Luz'
Mussaenda erythrophylla 'Doña Trining'
Mussaenda philippica 'Doña Aurora'
Mussaenda x 'Queen Sirikit'
Nerium oleander
Pittosporum tobira
Polyscias filicifolia 'Golden Prince'
Polyscias guilfoylei
Schefflera arboricola
Vitex trifolia *
Vitex trifolia variegata *

Small Trees:
Anacardium occidentale *
Brexia madagascariensis *
Brugmansia x candida
Caesalpinia pulcherrima
Caesalpinia pulcherrima 'Compton' *
Caesalpinia pulcherrima forma flava
Callistemon citrinus
Callistemon rigidus
Ceratonia siliqua *
Clusia rosea
Coccoloba uvifera
Coffea arabica
Coffea robusta *
Eriobotrya japonica *
Euphorbia cotinifolia
Hibiscus arnottianus subsp. punaluuensis
Hibiscus tiliaceus var. sterile
Hibiscus waimeae *
Myoporum sandwicense
Pandanus tectorius
Pittosporum hosmeri
Podocarpus elatus
Psidium cattleianum
Psidium cattleianum forma lucidum
Psydrax odorata
Schinus terebinthifolius
Senna surratensis
Tabebuia berteroi
Thevetia peruviana

Medium Trees:
Acacia confusa
Bucida buceras
Cassia x nealiae
Conocarpus erectus
Cordia subcordata
Cupressus sempervirens
Elaeodendron orientale
Ficus elastica *
Ficus elastica 'Decora' *
Ficus lyrata
Ficus rubiginosa
Filicium decipiens
Melaleuca quinquenervia
Metrosideros polymorpha
Mimusops caffra *
Nageia falcatus
Olea europaea subsp. europaea
Podocarpus macrophyllus
Schefflera actinophylla
Tabebuia heterophylla
Thespesia populnea

Large Trees:
Agathis robusta *
Agathis vitiensis *
Araucaria columnaris
Araucaria heterophylla *
Calophyllum inophyllum
Casuarina equisetifolia
Casuarina glauca *
Casuarina stricta *
Erythrina variegata
Erythrina variegata 'Tropic Coral'
Eucalyptus citriodora
Eucalyptus deglupta
Ficus macrophylla
Ficus microcarpa

Appendix E: House Plants

While no ornamental plant has evolved in an indoor environment, many decorative species may be brought inside with satisfactory results. Major sources of failure are low light intensity, lack of water, overwatering, or poor drainage. A weak, stringy-looking plant is often an indication of one or all three of these situations. House plant growers must also realize that a weakened plant becomes fair game for insects and other plant problems.

House plants fare better when a few recommendations are followed:

- Always remove excess water from a plant saucer; a plastic meat baster is useful;
- water when soil surface is dry; do not keep a watering schedule that might result in under- or overwatering;
- fertilize plants regularly and strictly according to manufacturer's instructions—there are many satisfactory fertilizers available;
- wash foliage at least once a week to remove dust;
- if possible, keep three specimens of each species, using one indoors for a week while the other two are outside; rotate on a regular basis;
- if maintaining more than one specimen of a desired house plant is not practical, give the plant a quarter turn each week to prevent one-sided growth (unless its decorative function calls for that shape, such as an espalier).

The following list is basic and may be augmented according to available light and the determination and talents of the indoor gardener. Those species marked with an asterisk (*) are not covered in the text but may be found in the list of suggested readings.

Small Plants:
Aglaonema commutatum
Aglaonema commutatum cultivars
Aglaonema modestum *
Aglaonema nitidum
Aglaonema nitidum cultivars
Anthurium spp. *
Anthurium 'Tropic Fire' and other dwarf hybrids *
Asparagus densiflorus 'Myers'
Asparagus densiflorus 'Sprengeri'
Aspidistra elatior *
Calathea spp.
Chloranthus inconspicuus *
Chlorophytum comosum
Codiaeum variegatum
Cordyline fruticosa cultivars
Dieffenbachia amoena
Dieffenbachia maculata
Dieffenbachia maculata cultivars
Dracaena godseffiana *
Dracaena sanderiana *
Epipremnum pinnatum 'Aureum'
Ficus deltoidea
Hoya spp. *
Peperomia obtusifolia
Peperomia spp. and cultivars *
Philodendron scandens forma micans
Philodendron spp. *
Pilea cadieri
Pilea depressa
Pilea involucrata *
Pilea mollis 'Moon Valley' *
Pilea nummularifolia
Pilea serpyllacea
Ruellia squarrosa
Ruellia makoyana
Saintpaulia ionantha cultivars *
Sansevieria trifasciata
Sansevieria trifasciata cultivars
Sansevieria spp. *
Spathicarpa sagittifolia *
Spathiphyllum spp. and cultivars
Syngonium auritum
Syngonium podophyllum
Syngonium podophyllum 'White Butterfly'

Large Plants:
Aralia elegantissima *
Araucaria columnaris
Araucaria heterophylla *
Clusia rosea
Coffea arabica
Dracaena fragrans *
Dracaena fragrans 'Warnecki'
Dracaena marginata
Dracaena marginata cultivars *
Ficus benjamina
Ficus benjamina 'Variegata' *
Ficus buchannani *
Ficus celebensis *
Ficus elastica *
Ficus elastica 'Decora' *
Ficus lyrata
Ficus macrophylla
Ficus microcarpa
Ficus microcarpa 'Variegata' *
Ficus rubiginosa
Grevillea robusta *
Leea guineensis
Monstera deliciosa
Nageia falcatus
Nolina recurvata
Podocarpus elatus
Podocarpus macrophyllus
Philodendron spp. *
Polyscias fruticosa
Polyscias guilfoylei 'Crispa'
Polyscias filicifolia 'Golden Prince'
Polyscias scutellaria 'Balfouriana'
Psidium cattleianum
Psidium cattleianum forma lucidum
Schefflera actinophylla
Schefflera arboricola

Bromeliads:
(Note: There is a wide range of bromeliads available to the house planter. Some of these are listed here.)
Aechmea fasciata
Cryptanthus spp.
Neoregelia carolinae and cultivars
Tillandsia cyanea
Tillandsia spp.
Vriesia hieroglyphica
Vriesia splendens *

Ferns:
(Note: There is a wide range of ferns available for use as house plants. Some of these are listed here.)
Adiantum capillus-veneris
Adiantum spp. and cultivars *
Asplenium nidus
Davallia fejeensis
Davallia spp. *
Microsorum punctatum 'Cristatum'
Nephrolepis exaltata
Nephrolepis exaltata cultivars
Nephrolepis falcata 'Furcans'

Bamboos:
(Note: Most species of bamboo are adaptable to use in a well-lighted interior. See text.)

Cycads:
Cycas revoluta
Zamia furfuracea

Palms:
Chamaedorea bachypoda
Chamaedorea cataractarum
Chamaedorea elegans *
Chamaedorea glaucifolia *
Chamaedorea graminifolia *
Chamaedorea metallica
Chamaedorea microspadix
Chamaedorea siefrizii
Coccothrinax argentea
Cyrtostachys renda
Dypsis lutescens
Howea belmoreana *
Howea forsteriana
Licuala grandis
Phoenix robelenii
Pinanga kuhlii
Ptychosperma macarthuri
Rhapis excelsa
Thrinax parviflora

Appendix F: Lanai Plants

The urge to mitigate the environment of high-rise living finds expression in efforts to grow plants on a lanai or balcony. This urge meets with varying results, from surprisingly satisfactory to disastrous. Given time, patience, and a modicum of information, the high-rise gardener can overcome the negative presence of strong winds, drought, heat, and the problems associated with the daily shift of sunlight from full sun to full shade. Recognition must be given to the inevitable need for some trials, errors, and adjustments and to the satisfaction of success.

Following are a few suggestions for enhancing the chances of success for lanai plants:

- Wash foliage frequently (at least once a week) to remove accumulated dust;
- water whenever the soil surface is dry; planting media on a lanai exposed to hot afternoon sun and strong winds will dry out rapidly; avoid a schedule for watering that may fail to supply water when needed or, on the other hand, result in overwatering;
- avoid letting water stand in saucers under pots for more than a few minutes; remove excess water (a plastic meat baster is good for this purpose);
- fertilize regularly; there are many satisfactory and easily applied fertilizers available;
- give potted plants a quarter turn every week to expose every part of the plant to maximum light; this will prevent "one-sided" plants;
- before beginning your lanai gardening, check with your building manager or maintenance supervisor to find out what the rules are, what weight your lanai can safely bear, and the means of disposing of runoff water.

The following table basically lists plants with small, leathery, wind-resistant foliage, combined with an ability to tolerate brief, reasonable drought. Experience with a specific environment enables the list to be greatly expanded to include other species attuned to those conditions. Those species marked with an asterisk (*) are not included in the text but are found in the list of suggested readings.

Small and Medium Plants:
Adenium multiflorum (P) *
Adenium obesum (P)
Adenium swazicum (P) *
Aloe vera
Aloe spp. *
Asparagus densiflorus 'Myers'
Asparagus densiflorus 'Sprengeri'
Carissa macrocarpa 'Boxwood Beauty' *
Chlorophytum comosum
Clusea 'Dwarf' *
Crassula ovata
Crassula ovata cultivars *
Ficus 'Green Island' *
Graptopetalum paraguariensis *
Hoya carnosa *
Kalanchoe spp. *
Ophiopogon jaburan 'vittatus'
Ophiopogon japonicus
Ophiopogon japonicus 'Compactus'
Pedilanthus spp. (P) (S) *
Peperomia obtusifolia
Pittosporum tobira 'Wheeler's Dwarf'
Portulacaria afra *
Portulacaria afra cultivars *
Rhaphiolepis indica
Rhaphiolepis umbellata var. integerrima
Sansevieria trifsciata
Sansevieria trifasciata cultivars *
Sansevieria spp. *
Scaevola taccada
Schefflera arboricola
Tradescantia spathacea
Tradescantia spathacea 'Dwarf'
Tradescantia zebrina
Wedelia trilobata
Zygocactus truncatus *
Zygocactus truncatus cultivars *

Large Plants:
Carissa macrocarpa
Clusia rosea
Conocarpus erectus
Dracaena marginata
Dracaena marginata cultivars *
Euphorbia lactea (P)$^+$(S)$^+$ *
Euphorbia tirucalli (P)$^+$(S)$^+$ *
Ficus benjamina
Ficus benjamina 'Variegata' *
Ficus elastica *
Ficus elastica 'Decora' *
Ficus lyrata
Ficus macrophylla
Ficus microcarpa
Ficus microcarpa var. crassifolia
Ficus rubiginosa
Kalanchoe beharensis
Leea guineensis
Ligustrum japonicum
Ligustrum japonicum 'Rotundifolium'
Murraya paniculata
Nageia falcatus
Nolina recurvata
Pittosporum tobira
Podocarpus elatus
Podocarpus macrophyllus
Polyscias filicifolia 'Golden Prince'
Polyscias fruticosa
Polyscias guilfoylei 'Crispa'
Polyscias scutellaria 'Balfouriana'
Psidium cattleianum
Psidium cattleianum forma lucidum
Schefflera actinophylla
Schinus terebinthifolius
Yucca elephantipes

Bromeliads:
(Note: Use the following on a partially shaded lanai.)
Aechmea fasciata
Cryptanthus spp.
Neoregelia carolinae and cultivars
Tillandsia cyanea
Vriesia hieroglyphica
Vriesia splendens *
(Note: Use the following on a sunny lanai.)
Aechmea blanchetiana
Ananas comosus var. variegatus
Neoregelia olens '696'
Tillandsia spp.
Vriesia imperialis

Ferns:
(Note: See House Plant fern list. These may be used on a sheltered lanai as above.)

Bamboos:
(Note: See House Plant bamboo list; these may be used as noted above.)

Cycads:
Cycas revoluta
Zamia furfuracea

Palms:
(Note: Those palms listed under House Plants may be used on a sheltered lanai where wind is greatly modified and in places where hot afternoon sun does not reach.)

Appendix G: The Garden at Night

The selection of plants for the nightscape is a completely individual matter. The choices are many and the opportunities for creativity are endless. Here are several considerations, however, that might be helpful in the creative process.

Plant foliage falls into two broad types: those that are translucent and those that are opaque. The former, when lighted from below or behind, glow with the quality of stained glass. This is experienced daily as the sun sets behind plants with translucent foliage or flowers. Opaque foliage, on the other hand, must have light falling on the leaf surface to elicit texture or color—although upward lighting of opaque foliage may result in interesting silhouettes or thrown shadow patterns. In certain situations, a combination of upward and direct light illuminating a plant produces a dramatic effect. The only rule is that there are no rules. Be flexible. Try many combinations until you find a lighting arrangement that fits your specific needs and taste.

In addition to foliage, trees with textured or colored bark or special branching patterns provide opportunities for adding both color and linear patterns to the nightscape. And, of course, many flowers—such as those of the Bougainvillea—are translucent and can provide major color accents during the flowering period.

Soft lighting is generally more pleasing and relaxing than strong lighting. Understatement is usually preferable to overstatement. In all cases, as in the selection and arrangement of plants for the nightscape, the handling of their illumination is a matter of personal taste. Always keep in mind that walking surfaces—especially steps and grade and directional changes—must be adequately lighted to provide safe passage. We suggest that you use a low-voltage system and call in an electrician before throwing the switch.

The following list is divided into three sections: (a) plants with translucent foliage; (b) plants with opaque foliage (those with white or silvery leaves are indicated by a plus [+], while those with strongly colored foliage are indicated by a double plus [++]); and (c) plants with strong bark color, texture, or strong branching patterns.

Species marked with an asterisk (*) are not listed in the text but are found in the list of suggested readings.

Plants with Translucent Foliage

Ground Covers:
Asparagus densiflorus 'Myers'
Asparagus densiflorus 'Springeri'
Chlorophytum comosum +

Small Shrubs:
Anthurium hookeri
Carludovica palmata *
Iresine herbstii ++

Medium Shrubs:
Crinum asiaticum
Dracaena marginata 'Variegata' + *
Graptophyllum pictum ++
Graptophyllum pictum 'Eldorado' ++
Heliconia indica 'Spectabilis' ++ *
Heliconia indica 'Striata' ++ *
Manihot esculentum 'Variegata' ++
Nandina domestica
Pseuderanthemum carruthersii var. atropurpureum ++
Pseuderanthemum carruthersii var. reticulatum ++
Pseuderanthemum carruthersii var. variegatum ++
Scaevola taccada
Xanthosoma robustum

Large Shrubs:
Acalypha godseffiana* ++
Acalypha godseffiana 'Heterophylla' ++
Acalypha wilkesiana ++
Acalypha wilkesiana 'Picotee' ++
Alocasia macrorrhizos
Breynia disticha 'Roseo-picta' ++
Cordyline fruticosa cultivars ++
Polyscias filicifolia 'Golden Prince' ++

Small Trees:
Euphorbia cotinifolia ++
Thevetia peruviana

Medium Trees:
Acacia confusa
Aleurites moluccana +
Bambusa spp. and other bamboo genera.
Pithecellobium dulce 'Variegata' +
Salix babylonica
Schinus molle
Tournefortia argentea +

Large Trees:
Artocarpus altilis
Barringtonia asiatica *
Hernandia nymphaeifolia *

Bromeliads:
Aechmea blanchetiana ++ (two color forms)
Neoregelia olens '646' ++
Vriesia hieroglyphica ++
Vriesia imperialis ++

Ferns:
Adiantum capillus-veneris
Adiantum capillus-veneris cultivars *
Adiantum spp. *
Angiopteris evecta
Angiopteris palmiformis *
Asplenium nidus
Cibotium chamissoi *
Cibotium glaucum
Cyathea cooperi
Microlepia strigosa
Microsorum punctatum 'Cristatum'
Nephrolepis cordifolia
Nephrolepis exaltata
Nephrolepis exaltata cultivars
Nephrolepis falcata 'Furcans'
Odontosoria chinensis
Phlebodium aureum
Phymatosorus grossus

Palms:
Caryota mitis
Caryota urens
Chamaedorea spp. All.
Dypsis lutescens
Licuala grandis
Licuala spinosa

Plants with Opaque Foliage

Ground Covers:
Heliotropium anomalum var. argenteum +
Hemigraphis alternata ++
Jaquemontia ovalifolia subsp. brasiliensis +
Pilea cadieri +
Rosmarinus officinalis +
Tradescantia pallida 'Purpurea' ++
Tradescantia zebrina ++
Vitex rotundifolia +

Small Shrubs:
Achyranthes splendens +
Aglaonema x 'Silver Queen' +
Aglaonema commutatum cultivars +
Aglaonema nitidum +
Aglaonema nitidum cultivars +
Artemisia mauiensis +
Crassula ovata +
Senecio cineraria +
Wikstroemia uva-ursi +

Medium Shrubs:
Agave attenuata +
Gossypium tomentosa +
Kalanchoe beharensis +
Leucophyllum frutescens +
Nototrichium sandwicense +
Olea europaea subsp. europaea 'Little Ollie' + *
Pittosporum tobira 'Variegata' +

Large Shrubs:
Bixa orellana ++
Calotropis gigantea +
Calotropis procera + *
Codaieum variegatum ++

Small Trees:
Heritiera littoralis +
Pandanus tectorius +
Tabebuia aurea +

Medium Trees:
Acacia koa +
Chrysophyllum oliviforme ++
Conocarpus erectus +
Ficus benjamina 'Variegata' + *
Ficus microcarpa 'Variegata' + *
Olea europaea subsp. *europaea* +

Bromeliads:
Aechmea fasciata +
Ananas comosus var. *variegatus* ++
Tillandsia spp. +
Vriesia aff. *regina* +

Cycads:
Encephalartos lehmannii +

Palms:
Dismarkia nobilis +
Brahea armata +

Coccothrinax argentata +
Copernicia prunifera +
Dypsis decaryi +
Latania loddigesii +
Pritchardia hillebrandii +
Pritchardia martii + *
Pritchardia pacifica

Plants with Strong Bark Color, Texture, or Strong Branching Patterns

Large Shrubs:
Eugenia uniflora (light-colored bark)

Small Trees:
Caesalpinia ferrea * (colored bark)
Callistemon citrinus (weeping habit)
Ficus carica * (white bark)
Guaiacum officinale (colored bark)
Nolina recurvata (trunk and base pattern)
Psidium cattleianum (reddish bark)
Psidium cattleianum forma *lucidum* (reddish bark)

Medium Trees:
Acacia confusa (light bark)
Bambusa spp. plus other bamboo genera (vertical accents)
Crescentia cujete (light bark, branching habit)

Cupressus sempervirens (vertical accent)
Litchi chinensis (light bark)
Melaleuca quinquenervia (white bark, bark texture)
Pimenta dioica (white bark)
Pimenta racemosa var. *racemosa* * (reddish bark)
Prosopis pallida (trunk pattern)
Pseudobombax ellipticum (colored and patterned bark)
Reynoldsia sandwicensis * (white bark)
Salix babylonica (weeping habit)
Schinus molle (weeping habit)

Large Trees:
Casuarina equisetifolia (weeping habit)
Enterolobium cyclocarpum (light bark)
Eucalyptus citriodora (white bark, weeping habit)
Eucalyptus deglupta (multicolored bark)
Ficus benghalensis (multiple trunks, aerial roots)
Ficus microcarpa (multiple trunks, aerial roots)

Palms:
(Note: The vertical, curving, or multiple trunks of all palms make interesting subjects for garden lighting.)

Glossary of Terms

Accent. The use of plant material to call attention to or create an emphasis in a design by means of its distinctive color, shape, size, or texture in contrast to plants used with it.

Arbor. An open structure upon which vines may be trained.

Aromatic. Plants that have a characteristic fragrance, especially when crushed.

Bonsai. The art of growing miniature container trees or shrubs by shaping and restricting the growth of their roots.

Bract. A modified leaf usually associated with a flower cluster, such as the colored bracts of the Poinsettia.

Bulbous. Enlarged to resemble a bulb in shape.

Calyx. A collective term for the sepals that form the outer whorl of the flower.

Compound leaf. A leaf usually composed of two or more parts or leaflets.

Conifers. A group of cone-bearing plants with needles, usually evergreen.

Corolla. A collective term for the flower petals.

Crown shaft. An apparent extension of the bole in some palms formed by the long, broad, overlapping, and sheathing bases of the leaves.

Cultivar. A horticulturally derived variety of a plant, as distinguished from a naturally occurring variety.

Deciduous. A plant that sheds all or nearly all of its leaves each year during the winter months or dry periods.

Endemic. Native to or confined naturally to a particular area or region.

Epiphytic. A plant normally found growing on another plant but not deriving its nourishment from it.

Espalier. A plant trained to grow flat against a wall or in one plane.

Evergreen. A plant that retains green and functional leaves throughout the year.

Genus. A botanical classification containing a group of species, all of which have certain structural characteristics that differ from those of other plants.

Glaucous. Covered with a bloom, giving a whitish or grayish color derived from a fine, waxy, powdery coating that is easily rubbed off.

Herbaceous. A plant that remains soft and succulent without extensive areas of lignified tissue; not woody.

Hybrids. The offsprings produced by breeding genetically dissimilar parent plants.

Indigenous. A native plant occurring or living naturally in an area but also naturally occurring in other areas.

Inflorescence. The flowering part of a plant and its mode of arrangement.

Interiorscaping. The design and arrangement of plants on the inside of buildings.

Leaflets. One of the segments or units of a compound leaf.

Nightscaping. Arranging and lighting a garden to be seen at night.

Palmate. A leaf with three or more lobes or leaflets radiating fanwise from a common basal point of attachment.

Perennial. A plant with a life span of three or more years.

Pergola. An open garden structure or arbor over which vines or other plants are trained.

Pinnate. A compound leaf with leaflets arranged on each side of a common axis, as with a feather.

Prostrate. Lying or growing flat on the ground.

Pubescent. Covered with or bearing short, soft hairs; hairy.

Raceme. An unbranched, indeterminate inflorescence in which the stalked flowers are arranged singly along a common main axis.

Rhizome. A rootlike, usually horizontal stem growing under or along the ground and sending up a succession of leaves or stems from its tip.

Rosette. Leaves arranged in a circular cluster from a crown, usually at or close to the ground.

Scalloped. A leaf edge with a series of rounded projections.

Semievergreen. A plant that may be evergreen or deciduous depending upon the climate.

Spadix. A thick or fleshy spike bearing minute flowers, usually surrounded or subtended by a spathe.

Spathe. A bract or leaf surrounding or subtending a flower cluster or spadix, sometimes colored.

Species. A basic unit in the botanical classification of plants.

Spore. The reproductive cell of lower or nonflowering plants such as fern.

Stamen. The male reproductive or pollen-bearing organ of flowering plants.

Succulent. A plant or plant part usually consisting of soft, fleshy, water-filled tissue.

Terrestrial. A plant living or growing in soil or on land.

Toothed. A leaf with sharp, sawtooth margins.

Twiner. A vine that coils or wraps around something.

Whorls. A circle of three or more leaves or other organs at one node.

Xeriscaping. Landscaping by grouping plants according to their water requirements and utilizing other water conserving measures.

Suggested Readings

Baensch, U., and U. Baensch. 1994. *Blooming Bromeliads*. Nassau, Bahamas: Tropic Beauty Publishers.

Bailey Hortorium (Cornell University). 1976. *Hortus Third*. New York: MacMillan.

Baldwin, R. E. 1990. *Hawaii's Poisonous Plants*. Hilo, HI: Petroglyph Press.

Bardi, P. M. 1964. *The Tropical Gardens of Robert Burle-Marx*. New York: Reinhold.

Beers, L., and J. Howie. 1992. *Growing Hibiscus*. Kenthurst, Australia: Kangaroo Press.

Belknap, J. P., M. Cazimero, N. E. Lewis, D. Peebles, and P. R. Weissich. 1982. *Majesty: The Exceptional Trees of Hawaii*. Honolulu: The Outdoor Circle.

Berry, R., and W. J. Kress. 1991. *Heliconia: An Identification Guide*. Washington, D.C.: Smithsonian Institution Press.

Bornhorst, H. L. 1996. *Growing Native Hawaiian Plants*. Honolulu: Bess Press.

Brenzel, K. N., Ed. 1995. *Sunset Western Garden Book*. Menlo Park, CA: Lane Magazine and Book Co.

Brown, B. F. 1995. *A Codiaeum Encyclopedia: Crotons of the World*. Valkaria, FL: Valkaria Tropical Garden.

———. 1994. *The Cordyline: King of Tropical Foliage*. Valkaria, FL: Valkaria Tropical Garden.

Clay, H. F., and J. C. Hubbard. 1977. *The Hawai'i Garden: Tropical Exotics*. Honolulu: University Press of Hawai'i.

———. 1977. *The Hawai'i Garden: Tropical Shrubs*. Honolulu: University Press of Hawai'i.

Denver Botanic Gardens. 1996. *Water Gardening*. New York: Knopf.

Dunk, G. 1994. *Ferns*. Sydney: Harper Collins.

Elbert, V. F., and G. A. Elbert. 1989. *Foliage Plants for Decorating Indoors*. Portland, OR: Timber Press.

Farrelly, D. 1997. *Book of Bamboo*. San Francisco: Sierra Club.

Graf, A. S. 1970. *Exotica 3*. E. Rutherford, NJ: Roehrs.

Griffiths, M. 1994. *The New Royal Horticultural Society Dictionary: Index of Garden Plants*. Portland, OR: Timber Press

Hodel, D. R. 1992. *Chamaedorea Palms*. Lawrence, KS: Allen Press.

Hoshizaki, B. J. 1976. *Fern Growers Manual*. New York: Knopf.

McDonald, E., ed. 1995. *Dry Climate Gardening with Succulents*. Huntington Botanical Garden. New York: Knopf.

Jones, D. L. 1993. *Cycads of the World*. Washington, D.C.: Smithsonian Institution Press.

———. 1995. *Palms throughout the World*. Washington, D.C.: Smithsonian Institution Press.

Krauss, B. H. 1993. *Plants in Hawaiian Culture*. Honolulu: University of Hawai'i Press.

Krempin, J. L. 1990. *Palms and Cycads around the World*. Sydney: Horowitz Grahame.

Leme, E. M. C., and L. C. Marigo. 1993. *Bromeliads in the Brazilian Wilderness*. Rio de Janeiro: Marigo Communiçao Visual.

Mabberley, D. J. 1990. *The Plant-Book*. Cambridge: Cambridge University Press.

Madulid, D. A. 1995. *A Pictorial Cyclopedia of Philippine Ornamental Plants*. Manila: Bookmark.

McCurrach, J. C. 1960. *Palms of the World*. New York: Harper & Brothers.

McCurrach, J. C., and A. C. Langlois. 1976. *Supplement to Palms of the World*. Gainesville: University Presses of Florida.

McDonald, M. A. 1978. *Ka Lei: The Leis of Hawaii*. Honolulu: Topgallant.

Meerow, A. W. 1992. *Betrock's Guide to Landscape Palms*. Cooper City, FL: Betrock Information Systems.

Moir, M. 1994. *Garden Watcher*. Honolulu: University of Hawai'i Press.

Morton, J. F. 1981. *500 Plants of South Florida*. Miami: Fairchild Tropical Garden.

———. 1982. *Plants Poisonous to People in Florida and Other Warm Areas*. Stuart, FL: Southern Printing.

Neal, M. C. 1965. *In Gardens of Hawaii*. Special Publication 50. Honolulu: Bishop Museum Press.

Northern, R. T. 1970. *Home Orchid Growing*. New York: Van Nostrand Reinhold.

Peterson, C. 1962. *Art of Flower Arrangement in Hawaii*. Honolulu: University of Hawai'i Press.

Philpotts, K. 1995. *Floral Traditions at the Honolulu Academy of Arts*. Los Angeles: Perpetua Press.

Pukui, M. K., and S. H. Elbert. 1986. *Hawaiian Dictionary*. Honolulu: University of Hawai'i Press.

Reynolds, G. W. 1950. *Aloes of South Africa*. Johannesburg: Aloes Book Fund.

———. 1966. *Aloes of Tropical Africa and Madagascar*. Mbabone, Swaziland: Aloes Book Fund.

Shimizu, H., and H. Takazawa. 1998. *New Tillandsia Handbook*. Tokyo: Japan Cactus Planting Co. Press.

Soonit, E. 1980. *Orchids of Asia.* Singapore: Times Books International.

Turgeon, A. J. 1996. *Turfgrass Management.* Englewood Cliffs, NJ: Prentice-Hall.

Uhl, N. W., and J. Dransfield. 1987. *Genera Palmarum.* Lawrence, KS: Allen Press.

Valier, K. 1995. *Ferns of Hawaii.* Honolulu: University of Hawai'i Press.

Wagner, W. L., D. R. Herbst, and S. H. Sohmer. 1999. *Manual of the Flowering Plants of Hawai'i, Revised Edition.* Honolulu: Bishop Museum Press and University of Hawai'i Press.

Weissich, P. R., D. Peebles, M. W. Cazimero, A. J. Ristoe, and L. L. Tom. 1991. *Majesty II: The Exceptional Trees of Hawaii.* Honolulu: The Outdoor Circle.

Warren, W. 1991. *The Tropical Garden.* New York: Thames and Hudson.

Index

'A'ali'i, 38
Abelia
   Glossy, 21
   x *grandiflora*, 21, 119
Acacia
   confusa, 57, 119, 122, 123
   koa, 57, 115, 123
Acalypha, 21
   godseffiana 'Heterophylla,' 21, 122
   hispida, 21
   Picotee, 35
   wilkesiana, 35, 122
   wilkesiana 'Picotee,' 35, 122
Achiote, 36
Achyranthes
   Hawaiian, 13
   splendens, 13, 115, 116, 117, 122
Acoelorraphe wrightii, 106, 118
Adenanthera pavonina, 57
Adenium obesum, 21, 116, 117, 121
Adiantum capillus-veneris, 92, 115, 120, 122
A'e, 68
'Ae'ae, 2
Aechmea
   blanchetiana, 89, 116, 117, 121, 122
   fasciata, 89, 120, 121, 123
African Fern Pine, 65
African Lily, 1
African Tulip Tree, 75
Agapanthus
   africanus. See *praecox* subsp. *orientalis*
   praecox subsp. *orientalis*, 1, 117
Agave
   attenuata, 22, 116, 117, 122
   Dragon Tree, 22
   Swan's Neck, 22
Aglaonema, 13
   commutatum, 13, 120
   commutatum 'Emerald Beauty,' 13, 120, 122
   commutatum tricolor 'Harlequin,' 13, 120, 122
   x 'José Rizal,' 13
   x 'Los Baños,' 13
   nitidum, 13, 120, 122
   nitidum 'Ernesto's Favorite,' 13, 120, 122
   x 'Silver Queen,' 13, 122
'Ahinahina, 14
Airplane Plant, 3
'Ākaha, 93
'Ākia, 19
'Alaea, 36
Alahe'e, 53
Alahe'e Haole, 42
Aleurites moluccana, 57, 115, 122
Alexandra Palm, 106
   King, 106
Alexandrian Laurel, 71
Algaroba, 67
Alibangbang, 58

Allamanda
   blanchetii, 22, 119
   Bush, 22
   cathartica, 77
   cathartica 'Hendersonii,' 77
   Common, 77
   Henderson, 77
   Purple, 22
   schotti, 22, 116, 119
Allspice Tree, 65
Aloalo Ko'ako'a, 41
Alocasia
   cucullata, 14
   macrorrhiza. See *macrorrhizos*
   macrorrhizos, 35, 122
   sanderiana, 14
Aloe
   Barbados, 1
   vera, 1, 116, 117, 121
Alpinia
   nutans. See *zerumbet*
   purpurata, 22
   zerumbet, 35
Alternanthera
   amoena. See *tenella*
   tenella, 1
Aluminum Plant, 8
Alyxia oliviformis, 77, 115
Amazon Blue, 17
American Rubber Plant, 8
Ananas comosus var. *variegatus*, 89, 116, 121, 123
Angel's Trumpet, 47
Angel Wing Jasmine, 81
Angiopteris evecta, 92, 122
Anisacanthus thurberi, 22, 116
Anthurium
   Bird's-Nest, 14
   hookeri, 14, 122
Antigonon leptopus, 77, 116
   'Album,' 77
'Ape, 34, 35
   Common, 34
Aptenia cordifolia, 1
Arabian Coffee, 49
Arabian Jasmine, 26
Arachis pintoi 'Golden Glory,' 2
Aralia
   False, 53
   Golden Prince, 44
   Ming, 30
   White-Edge Balfour, 44
Araucaria columnaris, 71, 119, 120
Archontophoenix alexandrae, 106
Ardisia crenata, 23
Areca Palm, 110
Arecastrum romanzoffianum. See *Syagrus romanzoffianum*
Argyreia nervosa, 77
Arrabidaea magnifica. See *Saritaea magnifica*
Artabotrys
   hexapetalus, 78
   uncinatus. See *hexapetalus*

Artemisia mauiensis, 14, 115, 122
Artillery Plant, 8
Artocarpus
   altilis, 71, 115, 122
   communis. See *altilis*
Asparagus
   densiflorus 'Myers,' 2, 116, 117, 120, 121, 122
   densiflorus 'Sprengeri,' 2, 116, 117, 120, 121, 122
   Fern, 78
   Foxtail, 2
   Myers, 2
   plumosus, 78
   setaceus. See *plumosus*
   Sprenger, 2
Asplenium nidus, 93, 115, 120, 122
Asystasia gangetica, 2, 116
Australian Tree Fern, 93
Autograph Tree, 48
'Awapuhi Ko'oko'o, 39
'Awapuhi Luheluhe, 35
'Awapuhi 'Ula'ula, 22
Azadirachta indica, 58, 116
Azalea, 32
   Indica, 32
   Mock, 21

Babylon Weeping Willow, 68
Baby Rubber Plant, 8
Baby's Tears, 8
Baby Wood Rose, 77
Bacopa monnieri, 2, 115
Bagauak, 37
Bagnit Vine, 88
Balsam, Zanzibar, 5
Bamboo
   Heavenly, 28
   Sacred, 28
Bamboo Cycad, 104
Bamboo Palm, 108, 113
   Hardy, 108
Bamboos, 101–103, 117, 120, 121, 122, 123
Bambusa, 122, 123
   malingensis, 101, 117
   multiplex 'Alphonse Karr,' 101, 117
   multiplex 'Silverstripe,' 101, 117
   textilis, 102
   vulgaris vittata, 102, 117
   wamin, 102
Banded Calathea, 98
Banyan, 73
   Chinese, 74
   Indian, 73
   Malayan, 74
Barbados Aloe, 1
Bastard Sandalwood, 52
Bauhinia
   binata, 58, 116
   x *blakeana*, 58
   galpinii, 78, 116
   monandra, 47
   Nasturtium, 78
   Pink, 47

punctata. See *galpinii*
Red, 78
tomentosa, 47, 116
variegata, 58
variegata 'Candida.' See *variegata*
Yellow, 47
Beach Heliotrope, 55
Beach Morning Glory, 6
Beach Naupaka, 33
Beach Vitex, 11
Beaucarnea, 52
Beaucarnea recurvata. See *Nolina recurvata*
Beaumontia jerdoniana. See *multiflora*
multiflora, 78
Beefsteak Plant, 16, 35
Begonia
Trailing, 79
Trailing Watermelon, 4
Beloperone guttata. See *Justicia brandeegeana*
Bengal Trumpet Vine, 87
White, 87
Benjamin Fig, 73
Benjamin Tree, 73
Be-Still Tree, 55
Bignonia, Purple, 86
Bird of Paradise, 33
Giant, 54
Bird's-Nest Fern, 93
Bird's Nest Sansevieria, 10
Bismarck Palm, 107
Bismarkia nobilis, 107, 123
Bixa orellana, 36, 123
Black Olive Tree, 59
Spiny, 47
Black-Eyed Stella Daylily, 5
Black Palm, 112
Blanchetiana
Lemon, 89
Orange, 89
Bleeding Heart Vine, 80
Blood Leaf, 16
Bloodwood, 75
Blue Butterfly Pea, 80
Blue Daze, 4
Blue Eranthemum, 24
Blue Flowered Clerodendrum, 37
Blue Ginger, 23
Blue Hesper Palm, 107
Blue Latan Palm, 111
Blue Plumbago, 30
Blue Sage, 24
Blue Trumpet Vine, 87
Blushing Palm, 109
Bombax, Pink, 67
Boston Fern, 94
Bottlebrush
Crimson, 48
Red, 48
Stiff, 48
Bottle Palm, 111
Bougainvillea
Brazil, 79
'California Gold,' 79
'Carmencita,' 79
'Mary Palmer,' 79
'Menehune,' 14

'Miss Manila,' 79
'Raspberry Ice,' 79
spectabilis, 79
'Temple Fire,' 14
Bowstring Hemp, 18
Bracket Plant, 3
Brahea armata, 107, 123
Brassaia, 68
Dwarf, 45
Brazil Bougainvillea, 79
Brazilian Rose, 61
Breadfruit, 71
Breynia
disticha 'Roseopicta,' 36, 119, 122
nivosa var. *roseo-picta.* See *disticha* 'Roseopicta'
Bromeliad, Striped Blushing, 90
Bromeliads, 89–92, 116, 117, 120, 121, 122, 123
Brown Pine, 67
Brugmansia x *candida,* 47, 119
Bucida
buceras, 59, 117, 119
molineti, 47
spinosa. See *molineti*
Burmese Rosewood, 75
Bush Allamanda, 22
Bush Thunbergia, 34
White, 34
Busy Lizzie, 5
Buttercup Tree, 61
Butterfly Gardenia, 45
Butterfly Pea, 80
Blue, 80
Buttonwood, Silver, 61

Cabbage Palm, 113
Caesalpinia
pulcherrima, 48, 116, 119
pulcherrima forma *flava,* 48, 116, 119
Cajeput Tree, 74
Calabash Tree, 62
Calathea, 120
albertii 'Emperor,' 96
Asian Beauty, 98
Banded, 98
burle-marxii 'Ice Blue,' 96
Emperor, 96
Fishtail, 97
Humilior, 98
Ice Blue, 96
leopardina, 96
majestica var. *roseo-lineata,* 97
majestica 'Princeps,' 96
makoyana, 97
orbifolia, 97
Princeps, 97
pseudoveitchiana, 97
Rose-Lined, 97
roseo-picta, 97
roseo-picta 'Asian Beauty,' 98
Rose-Streaked, 97
Round-Leaved, 97
Spotted, 96
variegata, 98
zebrina 'Humilior,' 98
Calico Plant, 36
California Pepper Tree, 68
Calliandra

haematocephala, 36
inaequilatera. See *haematocephala*
Callistemon
citrinus, 48, 119, 123
rigidus, 48, 119
Calophyllum inophyllum, 71, 115, 117, 119
Calotropis gigantea, 36, 116, 117, 119, 123
Candlenut Tree, 57
Canthium odoratum. See *Psydrax odoratum*
Cape Gardenia, 25
Cape Honeysuckle, 33
Yellow, 33
Cape Leadwort, 30
Cape Plumbago, 30
White, 30
Cardboard Palm, 106
Caribbee Royal Palm, 113
Caricature Plant, 25
Carissa
grandiflora. See *macrocarpa*
grandiflora var. *prostrata.* See *macrocarpa* 'Prostrata'
macrocarpa, 36, 116, 117, 119, 121
macrocarpa 'Prostrata,' 15
Carnauba Wax Palm, 109
Carpentaria acuminata, 107
Carpentaria Palm, 107
Caryota
mitis, 107, 122
urens, 107, 122
Cascabela thevetia. See *Thevetia peruviana*
Cascade Palm, 108
Cassava, Variegated, 28
Cassia
fistula, 59, 60
glauca. See *Senna surattensis*
grandis, 59
javanica, 59, 60
javanica x *fistula.* See x *nealiae*
x *nealiae,* 60, 119
x *nealiae* 'Lunalilo Yellow,' 60
x *nealiae* 'Nii Gold,' 60
x *nealiae* 'Queen's Hospital White,' 60
x *nealiae* 'Wilhelmina Tenney,' 60
Cassine orientalis. See *Elaeodendron orientale*
Casuarina equisetifolia, 72, 117, 119, 123
Cat Palm, 108
Cat's Claw Creeper, 82
Catharanthus roseus, 3
Cathedral Windows, 97
Cattley Guava, 52
Ceratozamia
hildae, 104
mexicana, 104
Ceriman, 83
Ceylon Wood Rose, 83
Chain of Love, 77
Chalice Vine, 86
Chamaedorea
brachypoda, 108, 120, 122
cataractarum, 108, 120, 122
metallica, 108, 120, 122
microspadix, 108, 120, 122
seifrizii, 108, 120, 122
Short-Stalked, 108, 122

Chambeyronia macrocarpa, 109
Chenille Plant, 21
Chinaberry, 65
Chinese Banyan, 74
Chinese Box, 42
Chinese Fan Palm, 111
Chinese Hat Plant, 41
Chinese Hedge Bamboo, 101
Chinese Hibiscus, 40
Chinese Silverstripe Hedge Bamboo, 101
Chinese Taro, 14
Chlorophytum comosum, 3, 120, 121, 122
Chorisia speciosa, 60
Christmas Berry Tree, 53
Christmas Flower, 39
Christmas Palm, 114
Christmas Vine, 85
Christ's Thorn, 15
Chrysalidocarpus lutescens. See *Dypsis lutescens*
Chrysophyllum oliviforme, 60, 123
Chusquea coronalis, 102
Cibotium glaucum, 93, 115, 122
Cigar Flower, 15
Circassian Bean, 57
Cissus discolor, 79
Citharexylum spinosum, 61
Clerodendrum
    Blue Flowered, 37
    buchanani var. *fallax*, 37
    inerme, 37, 117
    magnificum. See x *speciosum*
    myricoides 'Ugandense,' 37
    quadriloculare, 37
    Red, 80
    x *speciosum*, 79
    speciossimum. See *buchanani* var. *fallax*
    splendens, 80
    thomsonae, 79
    ugandense. See *myricoides* 'Ugandense'
    Vine, 79
Climbing Fig, 81
Climbing Ilang-Ilang, 78
Climbing Spray of Gold, 88
Clitoria ternatea, 80
Clover, Pink, 9
Clusia rosea, 48, 116, 117, 119, 120, 121
Clustering Fishtail Palm, 107
Coccoloba uvifera, 49, 116, 117, 119
Coccothrinax argentata, 109, 118, 120, 123
Cochlospermum
    vitifolium, 61
    vitifolium 'Plenum,' 61
Cock's Spur Coral Tree, 49
Coconut, Samoan, 109
Coconut Palm, 109
Cocos nucifera, 109, 115, 118
Codiaeum variegatum, 37, 119, 120, 123
Coffea arabica, 49, 119, 120
Coffee, 49
    Arabian, 49
Colvillea racemosa, 61
Common Allamanda, 77
Common 'Ape, 34
Common Coral Tree, 49
Common Gardenia, 25

Common Geranium, 17
Common Hibiscus, 40
Common Ironwood, 72
Common Oleander, 43
Common Panax, 44
Common Sword Fern, 94
Common Verbena, 11
Confederate Jasmine, 81, 88
Congea, 80
    griffithiana, 80
    velutina. See *griffithiana*
Conocarpus
    erectus, 61, 117, 119, 121, 123
    erectus sericeus. See *erectus*
Cook Pine, 71
Coontie, 106
Copernicia
    cerifera. See *prunifera*
    prunifera, 109, 123
Copey, 48
Copper Leaf, 35
Coprosma
    x *kirkii* 'Variegata,' 3
    repens 'Picturata,' 23
    Variegated Prostrate, 3
Coral Hibiscus, 41
Coral Plant, 32
Coral Shower, 59
Coral Tree
    Cock's Spur, 49
    Common, 49
    Indian, 72
Coral Vine, White, 77
Cordia
    sebestena, 49
    subcordata, 62, 115, 117, 119
Cordyline
    fruticosa, 38, 115, 119
    fruticosa 'Bob Alonzo,' 38, 119, 120, 122
    fruticosa 'Floozie,' 38, 119, 120, 122
    fruticosa 'Haole Girl,' 38, 119, 120, 122
    fruticosa 'Hawaiian Flag,' 38, 119, 120, 122
    fruticosa 'Iwao Shimizu,' 38, 119, 120, 122
    fruticosa 'Lau Kea,' 38, 119, 120, 122
    fruticosa 'Peter Buck,' 38, 119, 120, 122
    terminalis. See *fruticosa*
Coromandel, 2
Costa Rican Weeping Bamboo, 102
Cotton, Hawaiian, 25
Crane Flower, 33
Crape Myrtle
    Giant, 64
    Queen's, 64
Crassula
    argentea. See *ovata*
    ovata, 15, 116, 117, 121, 122
Creeping Charlie, 8
Creeping Fig, 81
Creeping Rosemary, 9
Crepe Jasmine, 45
Crescentia cujete, 62, 116, 123
Crested Fern, 94
Crimson Bottlebrush, 48
Crinum asiaticum, 23, 117, 122

Croton, 37
Crown Flower, 36
Crown of Thorns, 15
Cryptanthus, 120, 121
    'Ti,' 90
    zonatus, 90
Cryptostegia
    grandiflora. See *madagascariensis*
    madagascariensis, 80, 116, 117
Ctenanthe
    Burle-Marx, 98
    burle-marxii, 98
    oppenheimiana 'Tricolor,' 98
    pilosa 'Golden Mosaic,' 99
Cuban Mahogany, 69
Cuban Royal Palm, 113
Cup and Saucer, 41
    Yellow, 41
Cup of Gold, 86
Cuphea
    hyssopifolia, 3
    ignea, 15
Cupressus sempervirens, 62, 116, 119, 123
Curly Leaf Panax, 30
Cyathea
    australis. See *cooperi*
    cooperi, 93, 122
Cycad
    Bamboo, 104
    Karoo, 105
    Macdonnell Range, 105
    Small-Spined, 105
    Wooly, 105
Cycads, 104–106, 117, 120, 121, 123
Cycas
    circinalis, 104, 117
    revoluta, 104, 117, 120, 121
    Sago, 106
Cyperus
    alternifolius. See *involucratus*
    involucratus, 15
Cypress, Italian, 62
Cyrtostachys
    lakka. See *renda*
    renda, 110, 120

Dallas Fern, 94
Date Palm
    Dwarf, 112
    Pygmy, 112
Davallia fejeensis, 93, 120
Daylily
    Black-Eyed Stella, 5
    Golden Summer, 5
Delonix regia, 62, 116
Dendrocalamus
    brandesii, 102
    membranaceus, 103
Desert Honeysuckle, 22
Desert Rose, 21
Dewdrop
    Golden, 38
    White Golden, 38
Dichorisandra thyrsiflora, 23
Dictyosperma album, 110
Diffenbachia
    amoena, 23, 120
    amoena 'Tropic White,' 23

maculata, 24, 120
maculata 'Rudolph Roehrs,' 24, 120
maculata 'Superba,' 24, 120
maculata 'Uleryii,' 24, 120
Dioon spinulosum, 105
Dipladenia, 82
Dissotis
   plumosa. See *rotundifolia*
   rotundifolia, 3
Dodonaea viscosa, 38, 115, 116, 117, 119
Doña Aurora Mussaenda, 43
Doña Luz Mussaenda, 42
Doña Trining Mussaenda, 43
Doxantha unguis-cati. See *Macfadyena unguis-cati*
Dracaena
   deremensis 'Warneckii.' See *fragrans* 'Warneckii'
   fragrans 'Warneckii,' 38, 120
   marginata, 49, 120, 121
   Striped, 38
Dragon Tree
   Agave, 22
   Madagascar, 49
Dumb Cane, 23, 24
Dumb Plant, 23, 24
Duranta, 38
   erecta, 38, 119
   erecta 'Alba,' 38
   repens. See *erecta*
Dusty Miller, 18
Dwarf Brassaia, 45
Dwarf Date Palm, 112
Dwarf Geometry Tree, 47
Dwarf Ixora, Thai, 16
Dwarf Jamaican Heliconia, 16
Dwarf Lilyturf, 7
Dwarf Mondo Grass, 7
Dwarf Pittosporum, Wheeler's, 18
Dwarf Poinciana, 48
   Yellow, 48
Dwarf Pomegranate, 31
Dwarf Rosemary, 9
Dwarf Striped Bamboo, 103
Dypsis
   decaryi, 110, 123
   lutescens, 110, 120, 122

Earpod, 72
Easter Lily Vine, 78
'Ekaha, 93
Elaeodendron orientale, 63, 119
Elatostema repens, 4
Eldorado, 25
Elephant's Ear, 35, 72
Encephalartos
   lehmannii, 105, 118, 123
   villosus, 105
Enterolobium cyclocarpum, 72, 116, 123
Epipremnum pinnatum 'Aureum,' 81, 120
Eranthemum
   Blue, 24
   False, 31
   pulchellum, 24
   Purple False, 31
   Variegated False, 31
   Yellow-Veined False, 31
Erythea armata. See *Brahea armata*

Erythrina
   crista-galli, 49
   sandwicensis, 63, 115, 116
   Tall, 72
   variegata, 72, 119
   variegata orientalis. See *variegata*
   variegata 'Tropic Coral,' 72, 119
Etlingera elatior, 39
Eucalyptus
   citriodora, 73, 119, 123
   deglupta, 73, 119, 123
Eugenia uniflora, 39, 119, 123
Euphorbia
   cotinifolia, 50, 119, 122
   leucocephala, 39
   milii, 15, 116
   pulcherrima, 39
Euryops, 24
   pectinatis. See *pectinatis* × *E. abrotanifolius*
   pectinatis × *E. abrotanifolius*, 24
Everglades Palm, 106
Evolvulus
   glomeratus 'Blue Daze.' See *glomeratus* subsp. *grandiflorus*
   glomeratus subsp. *grandiflorus*, 4
'Ewa Hinahina, 13

Fagraea berterana, 63
False Aralia, 53
False Eranthemum, 31
   Purple, 31
   Variegated, 31
   Yellow-Veined, 31
False Heather, 3
False Kamani, 70
False Olive, 63
False Wiliwili, 57
Fan Palm
   Chinese, 111
   Fiji, 112
   Mexican, 114
   Thurston, 113
Farfugium japonicum, 4
Feather Bush, 42
Felt Bush, 27
Fern, Asparagus, 78
Fern Pine, African, 65
Ferns, 92–95, 115, 120, 121, 122
Fern Tree, 64
Ficus
   benghalensis, 73, 123
   benjamina, 73, 120, 121
   deltoidea, 24, 120
   lyrata, 63, 119, 120, 121
   macrophylla, 74, 116, 117, 119, 120, 121
   microcarpa, 74, 116, 117, 119, 120, 121, 123
   microcarpa var. *crassifolia*, 39, 117, 119, 121
   pumila, 81
   pandurata. See *lyrata*
   retusa. See *microcarpa*
   rubiginosa, 63, 119, 120, 121
Fiddle Leaf Fig, 63
Fiddlewood, 61
Fig
   Benjamin, 73
   Climbing, 81

   Creepling, 81
   Fiddle Leaf, 63
   Mistletoe, 24
   Moreton Bay, 74
   Port Jackson, 63
   Rusty, 63
   Taiwan, 39
   Wax, 39
   Weeping, 73
Fiji Fan Palm, 112
Filicium decipiens, 64, 119
Fire Spike, 29
Firecracker Hibiscus, 42
Firecracker Plant, 15, 32
Firecracker Vine, 85
Fish Bone Fern, 94
Fishtail Calathea, 97
Fishtail Fern, 94
Fishtail Palm, 107
   Clustering, 107
   Miniature, 108
Five Fingers Syngonium, 86
Flame Vine, Mexican, 85
Flamingo Flower, 27
Flor de Niño, 39
Floss Silk Tree, 60
Fluffy Duffy, 94
Formosan Koa, 57
Fountain Palm, 111
Foxtail Asparagus, 2
Foxtail Palm, 114
Fringed Hibiscus, 41
Fringeleaf Ruellia, 10

Galphimia
   glauca. See *gracilis*
   gracilis, 25
Ganges Violet, 2
Garden Geranium, 17
Gardenia
   augusta, 25, 119
   augusta 'Radicans,' 16
   Butterfly, 45
   Cape, 25
   Common, 25
   jasminoides. See *augusta*
   jasminoides 'Radicans.' See *augusta* 'Radicans'
   Paper, 45
   Prostrate, 16
   Tahitian, 40
   taitensis, 40, 117, 119
   Trailing, 16
Garden Verbena, 11
Garlic Vine, 83
Gazania
   rigens var. *leucolaena*, 4
   Trailing, 4
Geiger Tree, 49
Geometry Tree, 59
   Dwarf, 47
Geranium
   Common, 17
   Garden, 17
Giant Bird of Paradise, 54
Giant Crape Myrtle, 64
Giant Fern, 92
Giant Golden Bamboo, 102
Giant Lily, 23

Giant Milkweed, 36
Giant Taro, 35
Giant Yucca, 55
Ginger
   Blue, 23
   Red, 22
   Shell, 35
   Torch, 39
Glossy Abelia, 21
Gold-Banded Sansevieria, 10, 18
Golden Bamboo, Giant, 102
Golden Dewdrop, 38
   White, 38
Golden Flame Honeysuckle, 82
Golden Glory, 2
Golden Mosaic, 99
Golden Polypody, 95
Golden Pothos, 81
Golden Prince Aralia, 44
Golden Shower, 59
Golden Summer Daylily, 5
Golden-Fruited Palm, 110
Gold Heliconia, 40
Gold Tree, 75
Gossypium tomentosum, 25, 115, 116, 122
Graptophyllum
   pictum, 25, 119, 122
   pictum 'Eldorado,' 25, 119, 122
Grass, Mondo, 7
Green Ti, 38
Grewia occidentalis, 40
Guaiacum officinale, 50, 117, 123
Guava
   Cattley, 52
   Purple Strawberry, 52
   Yellow Strawberry, 53
Gum
   Lemon-Scented, 73
   Mindanao, 73

Hala, 52
Hall's Japanese Honeysuckle, 82
Hāpu'u, 93
Hardy Bamboo Palm, 108
Hare's-Foot Fern, Lacy, 93
Harpullia pendula, 50
Hau, 51
Hawaiian Achyranthes, 13
Hawaiian Cotton, 25
Hawaiian Tree Fern, 93
Hawthorn
   Indian, 32
   Yeddo, 32
Heart Leaf Philodendron, 84
Hearts and Flowers, 1
Heather
   False, 3
   Mexican, 3
Heavenly Bamboo, 28
Hedge Bamboo
   Chinese, 101
   Chinese Silverstripe, 101
Heliconia
   caribaea, 40
   Dwarf Jamaican, 16
   Gold, 40
   humilis 'Dwarf Jamaican.' See
     stricta 'Dwarf Jamaican'
   Parrot's, 16

Parrot's Beak, 26
psittacorum, 16
rostrata, 26
stricta 'Dwarf Jamaican,' 16
Heliotrope, Beach, 55
Heliotropium anomalum var. *argenteum*,
   4, 115, 116, 117, 122
Hemerocallis
   aurantiaca, 5
   'Black-Eyed Stella,' 5
Hemigraphis
   alternata, 5, 122
   alternata 'Exotica,' 5
   colorata. See *alternata*
   exotica. See *alternata* 'Exotica'
Hen's Eyes, 23
Henderson Allamanda, 77
Heritiera littoralis, 50, 117, 122, 123
Hibiscus
   arnottianus subsp. *punaluuensis*,
     50, 115, 119
   Chinese, 40
   Common, 40
   Coral, 41
   Firecracker, 42
   Fringed, 41
   Red, 40
   rosa-sinensis, 40, 119
   rosa-sinensis 'Higa Yellow,' 40, 119
   rosa-sinensis 'Lemon Chiffon,' 40,
     119
   rosa-sinensis 'Madam Pele,' 40, 119
   rosa-sinensis 'Nii Yellow,' 40, 119
   rosa-sinensis 'Snowflake,' 40, 119
   schizopetalus, 41, 119
   schizopetalus 'Butterfly White,' 41,
     119
   schizopetalus 'Itsy Bitsy Pink,' 41,
     119
   schizopetalus 'Pink Butterfly,' 41,
     119
   tiliaceus, 51, 115, 117
   tiliaceus var. *sterile*, 51, 119
   Tree, 51
Hierba Mala, 50
Hilo Holly, 23
Hinahina, 4, 14
Hispaniolan Rosy Trumpet Tree, 54
Hi'ui'a, 94
Hō'awa, 52
Holly
   Hilo, 23
   Singapore, 17
   West Indian, 41
Hollywood Juniper, 41
Holmskioldia
   sanguinea, 41, 119
   sanguinea 'Citrina,' 41, 119
Honeysuckle
   Cape, 33
   Desert, 22
   Golden Flame, 82
   Hall's Japanese, 82
   Japanese, 82
   Pink, 82
   Purple Japanese, 82
   Yellow Cape, 33
Hong Kong Orchid Tree, 58
Honohono, 11

Hottentot's Head, 105
Howea forsteriana, 110, 120
Huapala, 85
Hula Bamboo, 103
Huluhulu, 25
Hurricane Palm, 110
Hyophorbe lagenicaulis, 111, 118

I'iwi Haole, 33
Ilang-Ilang, Climbing, 78
'Ilie'e, 9
'Ilima, 33
'Ilima Papa, 10
Impatiens, 5
   hawkeri, 5
   New Guinea, 5
   wallerana, 5
Imperial Vriesia, 92
Indian Banyan, 73
Indian Coral Tree, 72
Indian Hawthorn, 32
Indian Laburnum, 59
Indian Laurel, 74
Indian Mulberry, 51
Indica Azalea, 32
Ipomoea
   horsfalliae, 81
   pes-caprae subsp. *brasiliensis*, 6,
     115, 117
Ipu Kula, 86
Iresine herbstii, 16, 122
Iris, Walking, 7
Ironwood, Common, 72
Italian Cypress, 62
Ivory Cane Palm, 112
'Iwa 'Iwa, 92
Ixora, 26
   casei 'Super King,' 26, 119
   chinensis 'Nora Grant,' 26, 119
   coccinea, 26, 119
   Nora Grant, 26
   Red, 26
   Super King, 26
   'Thai Dwarf,' 16

Jaburan Lilyturf, 7
Jacaranda, 64
   acutifolia. See *mimosifolia*
   mimosifolia, 64
Jacobinia, 27
Jacob's Coat, 35
Jade Plant, 15
   Large, 15
Jade Tree, 15
Jade Vine, 86
   Red, 83
   Scarlet, 83
Jaggery Palm, 107
Jamaican Heliconia, Dwarf, 16
Jamaican Thatch Palm, 114
Japanese Garden Juniper, 6
Japanese Honeysuckle, 82
   Hall's, 82
   Purple, 82
Japanese Pittosporum, 44
   Variegated, 30
Japanese Privet, 42
Japanese Sago Palm, 104
Japanese Yew, 67

Jaquemontia ovalifolia subsp.
   sandwicensis, 6, 115, 116, 117, 122
Jasmine
   Angel Wing, 81
   Arabian, 26
   Confederate, 81, 88
   Crepe, 45
   Madagascar, 83
   Star, 81, 88
Jasminum
   laurifolium forma *nitidum*, 81
   multiflorum, 81
   nitidum. See *laurifolium* forma
     nitidum
   sambac, 26, 116
Jatropha
   integerrima, 51
   Rose-Flowered, 51
Joyweed, 1
Jucaro, 59
Juniper
   Hollywood, 41
   Japanese Garden, 6
   Pfitzer, 26
   Shore, 6
Juniperus
   chinensis 'Pfitzeriana,' 26
   chinensis 'Torulosa,' 41
   conferta. See *procumbens*
   procumbens, 6, 117
Justicia
   brandeegeana, 27
   carnea, 27
   Red, 29

Kalamona, 54
Kalanchoe beharensis, 27, 116, 121, 122
Kamani, 71
   False, 70
Kamani Haole, 70
Kamehameha's Paddle, 90
Karoo Cycad, 105
Kentia Palm, 110
Kī, 38
Kiawe, 67
Kiele, 25
Kigelia
   africana, 64, 116
   pinnata. See *africana*
King Alexander Palm, 106
King of Bromeliads, 91
King Palm, 106
King Sago, 104
King's Crown, 27
King's Mantle, 34
Knotweed, 9
Koa, 57
   Formosan, 57
Koki'o Ke'oke'o, 50
Kokutan, 32
Kolokolo Kahakai, 11
Kolomana, 54
Kou, 62
Kou Haole, 49
Kris Plant, 14
Kūhiō Vine, 81
Kukui, 57
Kuluī, 29
Kumasaki Southern Yew, 67

Kupukupu, 94
Kupukupu Lau Li'i, 94

La'amia, 62
Laburnum, Indian, 59
Lace Fern, 95
Lacy Hare's-Foot Fern, 93
Lady Palm, 113
Lagerstroemia speciosa, 64
Lākana, 27
Lanalana, 78
Laniuma, 17
Lantana, 27
   Bush, 27
   camara, 27, 116, 117
   montevidensis, 6, 116, 117
   Trailing, 6
Large Jade Plant, 15
Large Staghorn Fern, 95
Latania loddigesii, 111, 123
Latan Palm, Blue, 111
Laua'e, 95
Laua'e Haole, 95
Lau'awa, 37
Laurel
   Alexandrian, 71
   Indian, 74
Laurel-Leaved Thunbergia, 87
Lavender Starflower, 40
Lechoso, 54
Leea, 41
   coccinea. See *guineensis*
   guineensis, 41, 119, 120, 121
Lehua Haole, 36
Lemon Blanchetiana, 89
Lemon-Scented Gum, 73
Leucophyllum frutescens, 28, 116, 119, 122
Licuala
   grandis, 111, 120, 122
   spinosa, 111, 122
Licuala Palm, 111
   Spiny, 111
Lignum Vitae, 50
Ligularia, 4
Ligularia tussilaginea. See *Farfugium japonicum*
Ligustrum
   japonicum, 42, 117, 119, 121
   japonicum 'Rotundifolium,' 42, 117, 119, 121
Lilac, Persian, 65
Liliko'i, 84
Lily
   African, 1
   Giant, 23
   Spider, 23
   White Rain, 11
   Zephyr, 11
Lilyturf, 7
   Dwarf, 7
   Jaburan, 7
   White, 7
Lipochaeta integrifolia, 6, 115, 116, 117
Lipstick Palm, 110
Lipstick Plant, 36
Liriope muscari, 7
Litchi chinensis, 64, 123
Livistona chinensis, 111

Lōkālia, 32
Lollipop Plant, 29
Looking Glass Tree, 50
Lonicera
   × *heckrottii*, 82
   japonica, 82
   japonica 'Atropurpurea,' 82
   japonica 'Halliana,' 82
Loulu Lelo, 112
Lumpy Noodle Bamboo, 102
Lychee, 64
Lysiloma thornberi, 42, 116

MacArthur Palm, 113
Macdonnell Range Cycad, 105
Macfadyena unguis-cati, 82, 116
Macrozamia macdonnellii, 105
Madagascan Three-Sided Palm, 110
Madagascar Dragon Tree, 49
Madagascar Jasmine, 83
Madagascar Olive, 65
Madagascar Periwinkle, 3
Madagascar Rubber Vine, 80
Magnificent Medinilla, 28
Mahogany
   Cuban, 68
   West Indian, 68
Maidenhair Fern, Southern, 92
Maile, 77
Maile Haole, 88
Maile-Scented Fern, 95
Malayan Banyan, 74
Maling Bamboo, 101
Malpighia coccigera, 17
Malvaviscus
   arobreus. See *penduliflorus*
   penduliflorus, 42, 119
Mandevilla, 82
   Alice Dupont, 82
   × *amabilis*, 82
   × *amabilis* 'Alice Dupont,' 82
   spendens. See × *amabilis*
   Yellow, 84
Mangrove, Silver Button, 61
Manihot esculenta 'Variegata,' 28, 122
Manila Palm, 114
Mansoa hymenaea, 83
Ma'o, 25
Maranta leuconeura var. *kerchoviana*, 99
Marantas, 96-99
Marsdenia floribunda, 83, 116, 117
Medinilla, 28
   magnifica, 28
   Magnificent, 28
Melaleuca
   leucadendra. See *quinquenervia*
   quinquenervia, 74, 116, 117, 119, 123
Melia azedarach, 65, 116, 117
Merremia tuberosa, 83
Merrill Palm, 114
Mesquite, 67
Metal Leaf, 5
Metallica, 108
Metrosideros polymorpha, 51, 115, 119
Mexican Creeper, 77
Mexican Fan Palm, 114
Mexican Flame Vine, 85
Mexican Heather, 3
Mexican Horncone, 104

Mexican Weeping Bamboo, 103
Mickey Mouse Plant, 29
Microlepia strigosa, 93, 115, 122
Microsorum
    punctatum 'Cristatum,' 94, 120, 122
    scolopendria. See *Phymatosorus*
        grossus
Milkweed, Giant, 36
Milo, 70
Mindanao Gum, 73
Ming Aralia, 30
Miniature Fishtail Palm, 108
Mirror Plant, Variegated, 23
Mistletoe Fig, 24
Moa, 95
Mock Azalea, 21
Mock Orange, 42
Monastery Bamboo, 103
Mondo Grass, 7
    Dwarf, 7
Money Tree, 49
Monkey Nut Palm, 114
Monkey Plant, 9
Monkeypod, 75
Monstera deliciosa, 83, 120
Moreton Bay Fig, 74
Morinda citrifolia, 51, 115, 116, 117
Morning Glory
    Beach, 6
    Woolly, 77
Moses in the Cradle, 10
Moss Rose, 9
Mother-in-Law's Tongue, 18
Mountain Apple, 69
Mountain Rose, 77
Mucuna
    bennettii. See *novoguineensis*
    novoguineensis, 83
Mulberry, Indian, 51
Mule's Foot Fern, 92
Murraya paniculata, 42, 119, 121
Mussaenda, 28
    Doña Aurora, 43
    x 'Doña Luz,' 42, 119
    Doña Trining, 43
    erythrophylla 'Doña Trining,' 43, 119
    flava. See *Pseudomussaenda flava*
    frondosa, 28, 119
    philippica 'Doña Aurora,' 43, 119
    x 'Queen Sirikit,' 43, 119
    Yellow, 31
Myers Asparagus, 2
Myoporum sandwicense, 52, 115, 116, 117, 119
Mysore Trumpet Vine, 88

Nageia falcatus, 65, 119, 120, 121
Naio, 52
Nānāhonua, 47
Nandina domestica, 28, 122
Nanten, 28
Narra, 75
Narrow Sword Fern, 94
Nasturtium Bauhinia, 78
Natal Plum, 36
    Prostrate, 15
Naupaka, 33
    Beach, 33
Naupaka Kahakai, 33

Neem Tree, 58
Nehe, 6
Neomarica gracilis, 7
Neoregelia
    carolinae forma *tricolor*, 90, 120, 121
    olens '696,' 90, 121, 122
Nephrolepis
    cordifolia, 94, 115, 122
    exaltata, 94, 115, 120, 122
    exaltata 'Bostoniensis,' 94, 120, 122
    exaltata 'Dallas,' 94, 120, 122
    exaltata 'Fluffy Duffy,' 94, 120, 122
    falcata 'Furcans,' 94, 120, 122
Nephthytis, 87
    White Butterfly, 87
Nerium oleander, 43, 116, 117, 119
Never-Never Plant, 98
New Guinea Impatiens, 5
Ni'ani'au, 94
Nim Tree, 58
Niu, 109
Nolina recurvata, 52, 116, 120, 121, 123
Noni, 51
Nora Grant Ixora, 26
Norfolk Island Pine, 71
Normanbya normanbyi, 112
Noronhia emarginata, 65, 117
Nototrichium sandwicense, 29, 115, 116, 117, 119, 122
Nova, 91

Ochna
    kirkii. See *thomasiana*
    thomasiana, 29
Octopus Tree, 68
Odontonema, 29
    cuspidatum, 29, 119
    strictum. See *cuspidatum*
Odontosoria chinensis, 95, 115, 122
'Ohai Ali'i, 48
'Ohe Kahiko, 103
'Ohe Nui, 102
'Ōhi'a 'Ai, 69
'Ōhi'a Lehua, 51
Olea
    europaea. See *europaea* subsp.
        europaea
    europaea subsp. *europaea*, 65, 116, 119, 123
Oleander, 43
    Common, 43
'Oliana, 43
Olive, 65
    Black, 59
    False, 63
    Madagascar, 65
    Spiny Black, 47
'Oliwa, 65
Operculina tuberosa. See *Merremia tuberosa*
Ophiopogon
    jaburan 'Vittatus,' 7, 121
    japonicus, 7, 121
    japonicus 'Compactus,' 7, 121
'Opiuma, White, 66
Orange Blanchetiana, 89
Orange-flowered Senecio, 85
Orange Trumpet Vine, 85
Orchid Tree, 58

    Hong Kong, 58
Orchid Vine, 86
Osmoxylon lineare, 17
Osteomeles anthyllidifolia, 29, 115, 116, 117, 119
Otatea acuminata aztecorum, 103, 117
Otocanthus coeruleus, 17
Oyster Plant, 10

Pachystachys lutea, 29, 119
Pagoda Flower, 37
Pahūpahū, 42
Pai Sang Bamboo, 103
Pala'ā, 95
Palai, 93
Palapalai, 93
Palms, 106–114, 115, 120, 121, 122, 123
Panax, 44
    Common, 44
    Curly Leaf, 30
    Parsley, 30
Pandanus
    odoratissimus. See *tectorius*
    tectorius, 52, 115, 117, 119, 123
Pānini 'Awa 'Awa, 1
Paperbark Tree, 74
Paper Gardenia, 45
Parasol Flower, 41
Parrot's Heliconia, 16
Parrot's Beak Heliconia, 26
Parsley Panax, 30
Pascuita, 39
Passiflora
    coccinea. See *vitifolia*
    edulis, 84
    vitifolia, 84
Passion Flower, Red, 84
Passion Fruit, 84
Pā'uohi'iaka, 6
Paurotis Palm, 106
Pea
    Blue Butterfly, 80
    Butterfly, 80
Peacock Plant, 97
Peanut
    Perennial, 2
    Pinto, 2
Pelargonium x *hortorum*, 17
Pellionia, 4
Pellionia daveauana. See *Elatostema repens*
Pentalinon lutea, 84
Pentas lanceolatus, 17
Peperomia obtusifolia, 8, 120, 121
Pepper Face, 8
Pepper Tree, 68
    California, 68
Peregrina, 51
Perennial Peanut, 2
Periwinkle, Madagascar, 3
Persian Lilac, 65
Peruvian Verbena, 11
Petrea volubilis, 84
Pfitzer Juniper, 26
Phaeomeria magnifica. See *Etlingera elatior*
Philodendron
    bipinnatifidum, 44
    Heart Leaf, 84

micans. See *scandens* forma *micans*
scandens forma *micans*, 84, 120
selloum. See *bipinnatifidum*
    Tree, 44
    Velvet Leaf, 84
Phlebodium aureum, 95, 122
Phoenix roebelenii, 112, 120
Phymatosorus grossus, 95, 122
Picotee Acalypha, 35
Pigeon Berry, 38
Pīkake, 26
Pilea
    cadierei, 8, 120, 122
    depressa, 8, 120
    nummulariifolia, 8, 120
    serpyllacea, 8, 120
Pilikai, 83
Pimenta dioica, 65, 116, 123
Pinanga kuhlii, 112, 120
Pine
    African Fern, 65
    Brown, 67
    Cook, 71
    Norfolk Island, 71
Pine-Tint Podocarpus, 67
Pineapple, Variegated, 89
Pink and White Shower, 59
Pink Bauhinia, 47
Pink Bombax, 67
Pink Clover, 9
Pink Honeysuckle, 82
Pink Quill, 90
Pink Sandpaper Vine, 80
Pink Shaving-Brush Tree, 67
Pink Shower, 59
Pink Tecoma, 69
Pink Trumpet Tree, 69
Pink Trumpet Vine, 85
Pinto Peanut, 2
Pitanga, 39
Pithecellobium dulce
    'Albo-variegatum.' See 'Variegata'
    'Variegata,' 66, 116, 122
Pittosporum
    hosmeri, 52, 115, 119
    Japanese, 44
    tobira, 44, 117, 119, 121
    tobira 'Variegata,' 30, 117, 119, 122
    tobira 'Wheeler's Dwarf,' 18, 117, 121
    tobira 'Wheelerii.' See *tobira* 'Wheeler's Dwarf'
    Variegated Japanese Pittosporum, 30
    Wheeler's Dwarf, 18
Platycerium superbum, 95
Pleioblastus viride striata, 103
Plumbago
    auriculata, 30, 116, 119
    auriculata 'Alba,' 30, 116
    Blue, 30
    Cape, 30
    capensis. See *auriculata*
    White Cape, 30
    Wild, 9
    zeylanica, 9, 115, 116
Plumeria, 66
    acuminata. See *rubra*
    acutifolia. See *rubra*
    obtusa, 66, 117
    rubra, 66, 117

rubra 'Common Yellow,' 66, 117
rubra 'Daisy Wilcox,' 66, 117
rubra 'Hilo Beauty,' 66, 117
rubra 'Kaneohe Sunburst,' 66, 117
rubra 'Kauka Wilder,' 66, 117
rubra 'Paul Weissich,' 66, 117
rubra 'Thornton Lilac,' 66, 117
Singapore, 66
Podocarpus
    elatus, 67, 119, 120, 121
    gracilior. See *Nageia falcatus*
    macrophyllus, 67, 119, 120, 121
    neriifolius. See *elatus*
    Pine-Tint, 67
Podranea ricasoliana, 85
Pōhinahina, 11
Pōhuehue, 6
Poinciana, 62
    Dwarf, 48
    Royal, 62
    Yellow Dwarf, 48
Poinsettia, 39
Poison Bulb, 23
Polypodium phymatoides. See *Phymatosorus grossus*
Polygonum capitatum, 9
Polyscias
    cumingiana 'Golden Prince.' See *filicifolia* 'Golden Prince'
    filicifolia 'Golden Prince,' 44, 119, 120, 121, 122
    fruticosa, 30, 119, 120, 121
    guilfoylei, 44, 119
    guilfoylei 'Crispa,' 30, 120, 121
    scutellaria 'Balfouriana,' 44, 120, 121
    scutellaria 'Marginata.' See *scutellaria* 'Balfouriana'
Pomegranate, 53
    Dwarf, 31
Pomeikalana, 53
Pony Tail, 52
Porana, 85
Porana paniculata. See *Poranopsis paniculata*
Poranopsis paniculata, 85
Portia Tree, 70
Port Jackson Fig, 63
Portulaca grandiflora, 9, 117
Pothos, Golden, 81
Powderpuff, Red, 36
Pride of Barbados, 48
Pride of India, 64, 65
Princeps Calathea, 96
Prince's Vine, 81
Princess Palm, 110
Pritchardia
    hillebrandii, 112, 118, 123
    pacifica, 112, 118, 123
    thurstonii, 113, 118
Privet
    Japanese, 42
    Round Leaf, 42
Prosopis pallida, 67, 116, 117, 123
Prostrate Coprosma, Variegated, 3
Prostrate Gardenia, 16
Prostrate Natal Plum, 15
Pseuderanthemun
    carruthersii, 31
    carruthersii var. *atropurpureum*, 31, 119, 122
    carruthersii var. *reticulatum*, 31, 119, 122
    carruthersii var. *variegatum*, 31, 119, 122
Pseudobombax ellipticum, 67, 123
Pseudocalymma alliaceum. See *Mansoa hymenaea*
Pseudogynoxys chenopodioides, 85, 116
Pseudomussaenda flava, 31
Psidium
    cattleianum, 52, 117, 119, 120, 121, 123
    cattleianum forma *lucidum*, 53, 117, 119, 120, 121, 123
    littorale var. *littorale.* See *cattleianum* forma *lucidum*
    littorale var. *longipes.* See *cattleianum*
Psilotum nudum, 95, 115
Psydrax odorata, 53, 115, 116, 117, 119
Pterocarpus indicus, 75
Ptychosperma macarthurii, 113, 120
Pua Kalaunu, 36
Pua Kenikeni, 63
Pua Kīkā, 15
Punica
    granatum, 53, 116
    granatum 'Nana,' 31
Purple Allamanda, 22
Purple Bignonia, 86
Purple False Eranthemum, 31
Purple Heart, 10
Purple Japanese Honeysuckle, 82
Purple Strawberry Guava, 52
Purple Tradescantia, 10
Purple Trumpet Tree, 69
Purple Wreath, 84
Pygmy Date Palm, 112
Pyrostegia
    ignea. See *venusta*
    venusta, 85

Queen Palm, 114
Queen Sago, 104
Queen's Crape Myrtle, 64
Queen Sirikit Mussaenda, 43
Quezonia, 37
Quisqualis indica, 85

Rabbit's-Foot Fern, 95
Rabbit Tracks, 99
Rainbow Shower, 60
Rain Lily, White, 11
Rain of Gold, 25
Rain Tree, 75
Rangoon Creeper, 85
Ravenala madagascariensis, 68
Red Bauhinia, 78
Red Bottlebrush, 48
Red Clerodendrum, 80
Red Ginger, 22
Red Hibiscus, 40
Red Hot Poker, 21
Red Ivy, 5
Red Ixora, 26
Red Jade Vine, 83
Red Justicia, 29
Red Palm, 110

Red Passion Flower, 84
Red Powderpuff, 36
Red Sandalwood, 57
Red Sealing Wax Palm, 110
Red Spurge, 50
Reed Palm, 108
Rex Begonia Vine, 79
Rhaphiolepis
   indica, 32, 117, 119, 121
   umbellata var. *integerrima*, 32, 117, 119, 121
   umbellata forma *ovata*. See *umbellata* var. *integerrima*
Rhapis excelsa, 113, 120
Rhododendron indicum, 32
Rondeletia odorata, 32
Rooftop Tillandsia, 91
Rose
   Desert, 21
   Mountain, 77
   Moss, 9
Rose-Flowered Jatropha, 51
Rose-Lined Calathea, 97
Rosemary
   Creeping, 9
   Dwarf, 9
Rose-Streaked Calathea, 97
Rosewood, Burmese, 75
Rosmarinus officinalis, 9, 116, 122
   'Prostratus,' 9
Rosy Trumpet Tree, 69
   Hispanolan, 54
Roth's Tillandsia, 91
Round Leaf Privet, 42
Round-Leaved Calathea, 97
Royal Palm, 113
   Caribbee, 113
   Cuban, 113
Roystonea
   oleracea, 113
   regia, 113
Rubber Plant
   American, 8
   Baby, 8
Ruellia
   ciliosa. See *squarrosa*
   Fringeleaf, 10
   makoyana, 9, 120
   squarrosa, 10, 120
Russelia equisetiformis, 32, 116, 117
Rusty Fig, 63

Sacred Bamboo, 28
Sago
   Cycas, 106
   King, 104
   Queen, 104
Sago Palm, 104
   Japanese, 104
Salix babylonica, 68, 122, 123
Samanea saman, 75
Samoan Coconut, 109
Sanchezia speciosa, 33, 119
Sandalwood
   Bastard, 52
   Red, 57
Sandpaper Vine, 84
   Pink, 80
Sansevieria
   Bird's Nest, 10
   Gold-Banded, 10, 18
   trifasciata, 18, 116, 117, 120, 121
   trifasciata 'Hahnii,' 10, 116, 120, 121
   trifasciata 'Laurentii,' 18, 116, 117, 120, 121
Sapindus saponaria, 68, 115
Saritaea magnifica, 86
Satinleaf, 60
Sausage Tree, 64
Scaevola
   sericea. See *taccada*
   taccada, 33, 115, 116, 117, 119, 121, 122
Scarlet Jade Vine, 83
Schefflera
   actinophylla, 68, 119, 120, 121
   arboricola, 45, 119, 120, 121
   elegantissima, 53
Schinus
   molle, 68, 116, 122, 123
   terebinthifolius, 53, 116, 119, 121
Schizostachys glaucifolium, 103
Scindapsus aureus. See *Epipremnum pinnatum* 'Aureum'
Scotch Attorney, 48
Scrambled Eggs, 54
Screwpine, 52
Sea Grape, 49
Sealing Wax Palm, 110
   Red, 110
Seminole Bread, 106
Senecio
   cineraria, 18, 116, 122
   confusus. See *Pseudogynoxys chenopodioides*
   Orange-flowered, 85
Senna surattensis, 54, 116, 117, 119
Sentry Palm, 110
Serissa japonica, 18
Shaving-Brush Tree, 67
   Pink, 67
Shell Flower, 35
Shell Ginger, 35
Shore Juniper, 6
Short-Stalked Chamaedorea, 108
Shower
   Coral, 59
   Golden, 59
   Lunalilo Yellow, 60
   Nii Gold, 60
   Pink, 59
   Pink and White, 59
   Queen's Hospital White, 60
   Rainbow, 60
   Wilhelmina Tenney, 60
   Yellow, 59
Shower of Orchids, 80
Shrimp Plant, 27
Sida fallax, 10, 33, 115, 116, 117
Silver Button Mangrove, 61
Silver Buttonwood, 61
Silver Evergreen, 13
Silver Palm, 109
Silver Trumpet Tree, 54
Silver Vase, 89
Silvery Vriesia, 92
Singapore Holly, 17
Singapore Plumeria, 66

'696,' 90
Small-Spined Cycad, 105
Snake Plant, 18
Snowbush, 36
Snow Creeper, 85
Soapberry, 68
Solandra
   guttata. See *maxima*
   maxima, 86
Sorcerer's Bush, 37
Southern Maidenhair Fern, 92
Spathiphyllum, 19, 120
   cannifolium. 19
   clevelandii. See *wallisii* 'Clevelandii'
   'McCoy,' 19
   'Sensation,' 19
   wallisii 'Clevelandii,' 19
Spathodea campanulata, 75
Sphenomeris chinensis. See *Odontosoria chinensis*
Spider Lily, 23
Spider Plant, 3
Spineless Yucca, 55
Spiny Black Olive, 47
Spiny Licuala Palm, 111
Spotted Calathea, 96
Spray of Gold, 25
   Climbing, 88
Sprenger Asparagus, 2
St. Thomas Tree, 47
Staghorn Fern, Large, 95
Stangeria eriopus, 105
Starflower, Lavender, 40
Star Jasmine, 81, 88
Stemmadenia
   galeottiana. See *litoralis*
   littoralis, 54
Stephanotis, 83
Stephanotis floribunda. See *Marsdenia floribunda*
Stiff Bottlebrush, 48
Stigmaphyllon
   floribundum, 86
   littorale. See *floribundum*
Strawberry Guava
   Purple, 52
   Yellow, 53
Strelitzia
   nicolai, 54
   reginae, 33
Striped Blushing Bromeliad, 90
Striped Dracaena, 38
Strongylodon macrobotrys, 86
Sundrops, 19
Super King Ixora, 26
Surinam Cherry, 39
Swan's Neck Agave, 22
Sweetpea Bush, 36
Swietenia mahogani, 69, 117
Swiss Cheese Plant, 83
Sword Fern
   Common, 94
   Narrow, 94
Syagrus romanzoffianum, 114
Syngonium
   auritum, 86, 120
   Five Fingers, 86
   podophyllum, 87, 120
   podophyllum 'White Butterfly,' 87, 120

Syzygium malaccense, 69, 115

Tabebuia
   aurea, 54, 123
   berteroi, 54, 119
   chrysantha. See *ochracea* subsp.
      neochrysantha
   donnell-smithii, 75
   heterophylla, 69, 119
   impetiginosa, 69
   ochracea subsp. *neochrysantha,* 75
   pallida. See *heterophylla*
   palmeri. See *impetiginosa*
   pentaphylla. See *rosea*
   rosea, 69
Tabernaemontana
   divaricata, 45
   divaricata 'Flore Pleno,' 45
Tahinu, 55
Tahitian Gardenia, 40
Taiwan Fig, 39
Tall Erythrina, 72
Taro
   Chinese, 14
   Giant, 35
Taro Vine, 81
Tecoma
   capensis, 33, 119
   capensis 'Aurea,' 33, 119
   Pink, 69
Tecomanthe dendrophila, 87
Temple Fire Bougainvillea, 14
Temple Tree, 66
Terminalia catappa, 70, 117
Texas Ranger, 28
Thai Dwarf Ixora, 16
Thespesia populnea, 70, 115, 117, 119
Thevetia peruviana, 55, 116, 117, 119, 122
Three-Seeded Mercury, 21
Three-Sided Palm, Madagascan, 110
Thrinax parviflora, 114, 118
Thunbergia
   Bush, 34
   erecta, 34, 119
   erecta 'Alba,' 34, 119
   grandiflora, 87
   grandiflora 'Alba,' 87
   Laurel-Leaved, 87
   laurifolia, 87
   mysorensis, 88
   White Bush, 34
Thurston Fan Palm, 113
Thyrsostachys siamensis, 103
Ti, 38
Tiare, 40
   Tahiti, 40
Tiger's Claw, 72
Tillandsia
   cyanea, 90, 120, 121, 123
   Rooftop, 91
   rothii, 91, 120, 121, 123
   Roth's, 91
   tectorum, 91, 120, 121, 123
   xerographica, 91, 116, 120, 121, 123
Tobira, 44
Toddy Palm, 107
Torch Ginger, 39
Tournefortia argentea, 55, 117, 122
Trachelospermum jasminoides, 88

Tradescantia
   pallida 'Purpurea,' 10, 116, 117, 122
   Purple, 10
   spathacea, 10, 116, 117, 121
   spathacea 'Dwarf,' 10, 116, 117, 121
   zebrina, 11, 121, 122
Trailing Begonia, 79
Trailing Gardenia, 16
Trailing Gazania, 4
Trailing Lantana, 6
Trailing Velvet Plant, 9
Trailing Watermelon Begonia, 4
Traveller's Palm, 68
Traveller's Tree, 68
Tree Fern,
   Australian, 93
   Hawaiian, 93
Tree Hibiscus, 51
Tree Philodendron, 44
Triangle Palm, 110
Tristellateia australasiae, 88
Tropical Almond, 70
Tropic Coral, 72
Trumpet Tree
   Hispaniolan Rosy, 54
   Pink, 69
   Purple, 69
   Rosy, 69
   Silver, 54
   Yellow, 76
Trumpet Vine
   Bengal, 87
   Blue, 87
   Mysore, 88
   Orange, 85
   Pink, 85
   White Bengal, 87
Tulip Tree, African, 75
Tulipwood, 50
Turk's Cap, 42
Turnera ulmifolia, 19, 116, 117

'Ulei, 29
'Ulu, 71
Umbrella Plant, 15
Umbrella Tree, 68
Urechites lutea. See *Pentalinon lutea*
Urn Plant, 89

Vada Tree, 73
Variegated Cassava, 28
Variegated False Eranthemum, 31
Variegated Japanese Pittosporum, 30
Variegated Mirror Plant, 23
Variegated Pineapple, 89
Variegated Prostrate Coprosma, 3
Veitchia merrillii, 114
Velvet Leaf, 27
Velvet Leaf Bamboo, 102
Velvet Leaf Philodendron, 84
Velvet Plant, Trailing, 9
Venus Hair Fern, 92
Verbena
   Common, 11
   Garden, 11
   x *hybrida,* 11, 116, 117
   Peruvian, 11
   peruviana. See x *hybrida*
Violet, Ganges, 2

Vitex
   Beach, 11
   rotundifolia, 11, 115, 116, 117, 122
Vriesia
   gigantea seidelii 'Nova,' 91
   hieroglyphica, 91, 120, 121, 122
   Imperial, 92
   imperialis, 92, 116, 121, 122
   aff. *regina,* 92, 123
   Silvery, 92

Waffle Plant, 5
Waiawī, 53
Waiawī 'Ula 'Ula, 52
Walking Iris, 7
Wamin Bamboo, 102
Wandering Jew, 11
Washingtonia Palm, 114
Washingtonia robusta, 114, 118
Washington Palm, 114
Water Hyssop, 2
Watermelon Begonia, Trailing, 4
Wax Fig, 39
Wax Palm, 109
   Carnauba, 109
Weaver's Bamboo, 102
Wedelia trilobata, 11, 117, 121
Weeping Bamboo
   Costa Rican, 102
   Mexican, 103
Weeping Fig, 73
Weeping Willow, 68
   Babylon, 68
West Indian Holly, 41
West Indian Mahogany, 69
Wheeler's Dwarf Pittosporum, 18
Whisk Fern, 95
White Bengal Trumpet Vine, 87
White Bush Thunbergia, 34
White Cape Plumbago, 30
White Coral Vine, 77
White Dragon, 7
White-Edge Balfour Aralia, 44
White Golden Dewdrop, 38
White Lilyturf, 7
White 'Opiuma,' 66
White Rain Lily, 11
Wikstroemia uva-ursi, 19, 115, 116, 117, 122
Wild Plumbago, 9
Wilelaiki, 53
Wiliwili, 63
   False, 57
Wiliwili Haole, 72
Willow
   Babylon Weeping, 68
   Weeping, 68
Wine Palm, 107
Wodyetia bifurcata, 114
Wood Rose, 83
   Baby, 77
   Ceylon, 83
Woolly Morning Glory, 77
Wooly Cycad, 105

Xanthosoma robustum, 34, 122
Xerographica, 91

Yeddo Hawthorn, 32

Yellow Alder, 19
Yellow Bauhinia, 47
Yellow Cape Honeysuckle, 33
Yellow Cup and Saucer, 41
Yellow Dwarf Poinciana, 48
Yellow Lollipop, 29
Yellow Mandevilla, 84
Yellow Mussaenda, 31
Yellow Shower, 59
Yellow Strawberry Guava, 53

Yellow Trumpet Tree, 76
Yellow-Veined False Eranthemum, 31
Yew
    Japanese, 67
    Kumasaki Southern, 67
Ylang Ylang, 78
Yucca
    elephantipes, 55, 116, 117, 121
    Giant, 55
    Spineless, 55

Zamia
    furfuracea, 106, 116, 118, 120, 121
    integrifolia, 106, 118
Zanzibar Balsam, 5
Zebra Plant, 90
Zephyr Lily, 11
Zephyranthes candida, 11

About the Authors

Fred D. Rauch, Ph.D., is emeritus professor of horticulture at the University of Hawai'i, where he served as extension specialist in horticulture for twenty-five years. His interest in ornamental plants began while studying for the B.S. degree in horticulture (with emphasis on landscape construction and maintenance) at Oregon State University. The study of tropical landscape plants was initiated through research and instruction as an assistant professor of horticulture at Mississippi State University. He has undertaken extensive research on the identification and use of tropical ornamental plants in the landscape, resulting in numerous publications. Among these is a comprehensive laboratory manual for use in teaching the tropical plant courses in the University of Hawai'i system. He has developed many training classes for the "Green Industry" in Hawai'i and was instrumental in the formation of the Landscape Industry Council of Hawai'i.

Paul R. Weissich, A.S.L.A., is currently a licensed landscape architect whose familiarity with tropical landscape species has been recognized professionally, resulting in numerous consultant assignments. He is also an active volunteer in several community areas. From 1957 to 1989, Weissich was director of the Honolulu Botanical Gardens, where he expanded the two-garden system from 50 to 650 acres, covering four sites of differing ecological situations. The plant collection was increased to a position of international recognition. Staff and budget were appropriately expanded and public education programs were developed, ranging from formal classes and demonstrations to field trips on O'ahu and the neighbor islands and to Central and South America, the Galapagos Islands, and New Zealand. A corps of volunteers was trained to provide free guided educational tours to the several garden sites, serving visitors of all ages. The planning and directing of all garden operations provided an ideal vehicle for acquiring knowledge of an extremely wide range of tropical species, many of which are threatened and endangered.